T0226377

SpringerBriefs in Astronomy

Series editors

Martin Ratcliffe
Valley Center, Kansas, USA

Wolfgang Hillebrandt
MPI für Astrophysik, Garching, Germany

Michael Inglis
SUNY Suffolk County Community College, Selden, New York, USA

David Weintraub
Vanderbilt University, Nashville, Tennessee, USA

SpringerBriefs in Astronomy are a series of slim high-quality publications encompassing the entire spectrum of Astronomy, Astrophysics, Astrophysical Cosmology, Planetary and Space Science, Astrobiology as well as History of Astronomy. Manuscripts for SpringerBriefs in Astronomy will be evaluated by Springer and by members of the Editorial Board. Proposals and other communication should be sent to your Publishing Editors at Springer.

Featuring compact volumes of 50 to 125 pages (approximately 20,000–45,000 words), Briefs are shorter than a conventional book but longer than a journal article. Thus Briefs serve as timely, concise tools for students, researchers, and professionals.

Typical texts for publication might include:

- A snapshot review of the current state of a hot or emerging field
- A concise introduction to core concepts that students must understand in order to make independent contributions
- An extended research report giving more details and discussion than is possible in a conventional journal article
- A manual describing underlying principles and best practices for an experimental technique
- An essay exploring new ideas within astronomy and related areas, or broader topics such as science and society

Briefs allow authors to present their ideas and readers to absorb them with minimal time investment.

Briefs will be published as part of Springer's eBook collection, with millions of readers worldwide. In addition, they will be available, just like other books, for individual print and electronic purchase.

Briefs are characterized by fast, global electronic dissemination, straightforward publishing agreements, easy-to-use manuscript preparation and formatting guidelines, and expedited production schedules. We aim for publication 8–12 weeks after acceptance.

More information about this series at http://www.springer.com/series/10090

Klara Anna Capova · Erik Persson
Tony Milligan · David Dunér
Editors

Astrobiology and Society in Europe Today

EUROPEAN COOPERATION
IN SCIENCE & TECHNOLOGY

Editors
Klara Anna Capova
Department of Anthropology
Durham University
Durham, UK

Tony Milligan
Department of Theology & Religious
 Studies
Kings College London
London, UK

Erik Persson
Department of Philosophy
Lund University
Lund, Sweden

David Dunér
Department of Arts and Cultural Sciences
 and Division of History of Ideas and
 Sciences
Lund University
Lund, Sweden

ISSN 2191-9100   ISSN 2191-9119 (electronic)
SpringerBriefs in Astronomy
ISBN 978-3-319-96264-1   ISBN 978-3-319-96265-8 (eBook)
https://doi.org/10.1007/978-3-319-96265-8

Library of Congress Control Number: 2018948604

© The Author(s), under exclusive license to Springer International Publishing AG, part of Springer Nature 2018
This work is subject to copyright. All rights are reserved by the Publisher, whether the whole or part of the material is concerned, specifically the rights of translation, reprinting, reuse of illustrations, recitation, broadcasting, reproduction on microfilms or in any other physical way, and transmission or information storage and retrieval, electronic adaptation, computer software, or by similar or dissimilar methodology now known or hereafter developed.
The use of general descriptive names, registered names, trademarks, service marks, etc. in this publication does not imply, even in the absence of a specific statement, that such names are exempt from the relevant protective laws and regulations and therefore free for general use.
The publisher, the authors, and the editors are safe to assume that the advice and information in this book are believed to be true and accurate at the date of publication. Neither the publisher nor the authors or the editors give a warranty, express or implied, with respect to the material contained herein or for any errors or omissions that may have been made. The publisher remains neutral with regard to jurisdictional claims in published maps and institutional affiliations.

This Springer imprint is published by the registered company Springer Nature Switzerland AG
The registered company address is: Gewerbestrasse 11, 6330 Cham, Switzerland

Acknowledgements

This publication is based upon the work from COST Action Origins and evolution of life on Earth and in the Universe (ORIGINS) supported by COST (European Cooperation in Science and Technology).

COST (European Cooperation in Science and Technology) is a funding agency for research and innovation networks. Our Actions help connect research initiatives across Europe and enable scientists to grow their ideas by sharing them with their peers. This boosts their research, career and innovation.

www.cost.eu

Funded by the Horizon 2020 Framework Programme of the European Union

The editors would like to acknowledge the support of COST Action TD1308 ORIGINS through the Short-Term Scientific Mission (STSM) fundings: STSM-TD1308-010216-070847, STSM-TD1308-121216-081651, STSM-TD1308-1604180-40883 and STSM-TD1308-150418-040842 and the support of the Pufendorf Institute for Advanced Studies at Lund University, Sweden, through the "A Plurality of Lives" research theme.

The editors would also like to thank all members of the international advisory board: Octavio Chon-Torres, Kathryn Denning, Steven Dick, Abhik Gupta, Sun Kwok, Lucas Mix Addy Pross, and Kai-Uwe Schrogl.

The editors would also like to thank all of the authors and researchers who contributed to the Astrobiology and Society White Paper. For their comments on the text, we extend our thanks to Julie Nováková (Charles University in Prague, Czech Republic), Caroline Dorn (University of Zurich, Switzerland), Vladimir Bozhilov (Sofia University, Bulgaria), and John Robert Brucato (Astrophysical Observatory of Arcetri, Italy).

About this Book

This book describes the state of astrobiology in Europe today and its relation to the European society at large. With contributions from authors in more than 20 countries and over 30 scientific institutions worldwide, the document illustrates the societal implications of astrobiology and the positive contribution that astrobiology can make to European society.

The book has two main objectives: 1. It recommends the establishment of a European Astrobiology Institute (EAI) as an answer to a series of challenges relating to astrobiology but also European research, education, and society at large. 2. It also acknowledges the societal implications of astrobiology, and thus the role of the social sciences and humanities in optimizing the positive contribution that astrobiology can make to the lives of the people of Europe and the challenges they face.

Astrobiology enjoys a great deal of interest among the public, probably more than most of the other fields of research. It also has implications for human life outside the laboratories and lecture halls. It has the potential of being a flagship of European cooperation in science. It provides an ideal ground for collaborative European projects which support the ethos of cooperating countries. Astrobiology is inherently multidisciplinary and based on collaboration between disciplines, universities, and countries. For Europe to take a leading role in this research, it is very important to have a stable structure that can coordinate research, research infrastructure, funding and relations to the surrounding society in an efficient way. The establishment of a EAI, as a consortium of institutions, will provide the perfect forum for such collaborative efforts and should be a key priority for European research institutions as well as the European astrobiology community and the EU. To have an active astrobiology research programme, coordinated and fostered by such an institute, will enhance the international standards of European space research and of European science in general.

The EAI would be able to promote astrobiology research, assist in the decision-making process of relevant European institutions, be involved in mission planning, engage in science dissemination, education and communication, as well as engaging in outreach and media work in a much more efficient way than individual research institutions. The EAI will act as a strong voice for the astro-biology community in dialogue with decision makers, funding agencies, the media, other stakeholders, and the general public. It will be proactive in the debate on important legal and ethical issues in astrobiology and space research.

Contents

Editors and Contributors

A. Anglés University of Hong Kong, Pokfulam, Hong Kong

J. Arnould Centre national d'études spatiales (National Centre for Space Studies), Paris, France

L. Billings Consultant to NASA's Astrobiology Program, California, USA

K. A. Capova Department of Anthropology, Durham University, Durham, UK

E. Chatzitheodoridis School of Mining and Metallurgical Engineering, National Technical University of Athens, Athens, Greece

L. Dartnell University of Westminster, London, UK

D. Dunér Lund University, Lund, Sweden

M. Gargaud Laboratoire d'Astrophysique de Bordeaux, Pessac, France

W. Geppert Physics Department (Fysikum), Stockholm University (Stockholms universitet), Stockholm, Sweden

E. Hemminger Evangelische Hochschule Rheinland-Westfalen-Lippe, Bochum, Germany

Z. Kaňuchová Astronomický ústav SAV (Astronomical Institute SAS), Tatranská Lomnica, Slovakia

A. Kereszturi Csillagászati ès Földtudományi Kutatóközpont (Research Centre for Astronomy and Earth Sciences), Budapest, Hungary

G. Kminek European Space Agency, Noordwijk, The Netherlands

P. Laine Jyväskylän Yliopisto, University of Jyväskylä, Jyväskylä, Finland

A. Losiak Polska Akademia Nauk (Polish Academy of Sciences), Warsaw, Poland

Z. Martins Instituto Superior Técnico, Universidade de Lisboa, Lisbon, Portugal

J. Martínez-Frías Instituto de Geociencias, IGEO (CSIC-UCM), Madrid, Spain

N. Mason School of Physical Science, The Open University, Milton Keynes, UK

A. Melin Malmö Universitet, Malmö, Sweden

T. Milligan Department of Theology and Religious Studies, King's College London, London, UK

P. T. Mitrikeski Institut za istraživanje i razvoj održivih ekosustava (Institute for Research and Development of Sustainable Ecosystems), Zagreb, Croatia

E. Nabulya Makerere University, Kampala, Uganda

L. Noack Freie Universität Berlin (Free University Berlin), Berlin, Germany

E. Persson Department of Philosophy, Lund University, Malmö, Sweden

S. Ramos Universidad Nacional Autónoma de México, Mexico, USA

K. Smith Clemson University, Clemson, USA

S. Tirard Université de Nantes, Nantes, France

M. Waltemathe Ruhr-Universität Bochum, Bochum, Germany

Abbreviations

AbGradE	Astrobiology Graduates in Europe
ASB	Astrobiology Society of Britain
ASSA	Astronomical Society of Southern Africa
COSPAR	Committee on Space Research
COST	Cooperation in Science and Technology
DAbG	German Astrobiology Society
EAC	European Astrobiology Campus
EAI	European Astrobiology Institute
EANA	European Astrobiology Network Association
EC	European Commission
ELSI	Earth-Life Science Institute
ESA	European Space Agency
ESO	European Southern Observatory
EU	European Union
FAST	Five Hundred Meter Aperture Spherical Telescope
IAGETH	International Association for Geoethics
IAU	International Astronomical Union
ICSU	International Council for Science
ISSI	International Space Science Institute
ISU	International Space University
JAXA	Japan Aerospace Exploration Agency
NAI	NASA Astrobiology Institute
NASA	National Aeronautics and Space Administration
NNA	Nordic Network of Astrobiology
OST	Outer Space Treaty
SFE	The French Astrobiology Society
SIA	Italian Society of Astrobiology
SOMA	Mexican Society of Astrobiology
STEM	Science, Technology, Engineering, Mathematics
STS	Science and Technology Studies

STSM	Short-Term Scientific Mission
TD	Trans Domain
UN	United Nations
UNAM	National Autonomous University of Mexico
UNESCO	UN Educational, Scientific and Cultural Organization
UNOOSA	United Nations Office for Outer Space Affairs
WG5	Working Group 5

List of Figures

Chapter 1
Introduction

T. Milligan, K. A. Capova, D. Dunér and E. Persson

Astrobiology is an expanding multidisciplinary scientific research field. It touches upon some very old questions about where we come from, about the possibilities for life, and about the prospects for humanity. As a scientific multidisciplinary field, it focuses upon the origin, evolution and future of life. Formal definitions of 'astrobiology' (including the definition used in the White Paper that follows) tend to cite these matters. Astrobiology does not, however, simply repeat familiar ways of looking at our origins and life's evolution and future. Its newness is genuinely new. An exploration of the practices of a contemporary astrobiologist and of someone like the ancient poet Lucretius, speculating about these same matters but separated by two millennia of time, would show many differences. The most obvious difference is that astrobiology, unlike even the most remarkable examples of early speculation, brings a scientific rigour to longstanding questions about life, allowing them to be posed in new ways, in the light of a large body of discoveries across a range of scientific disciplines.

For example, contemporary astrobiologists pose questions about life in the light of the discovery of thousands of planets in other solar systems, and of life on Earth that might survive and even thrive in extreme environments elsewhere, i.e. so-called extremophiles. Such discoveries also allow entirely new, and previously unimagined, questions to be posed: not only questions about such microbial forms of life and why they are capable of surviving in extreme environments, but also questions about how to distinguish between life and non-life in difficult cases, how to distinguish

T. Milligan (✉)
Department of Theology and Religious Studies, Room 3.42, Virginia Woolf Building, King's College London, 22 Kingsway, WC2B 6LE London, UK
e-mail: Anthony.milligan@kcl.ac.uk

K. A. Capova
Durham University, Durham, UK

D. Dunér · E. Persson
Lund University, Lund, Sweden

© The Author(s), under exclusive license to Springer International Publishing AG, part of Springer Nature 2018
K. A. Capova et al. (eds.), *Astrobiology and Society in Europe Today*, SpringerBriefs in Astronomy, https://doi.org/10.1007/978-3-319-96265-8_1

between living organisms and chrystalline structures whose growth may happen to share some features in common with living things. Discovery also allows us to pose new questions about how best to define life, and about whether or not a satisfactory definition of life is currently or ultimately possible, and in what sense such a definition might be necessary for certain forms of scientific enquiry.

Where Europe sits within this emerging research field, is a matter of some interest. While work on life is a feature of scientific research worldwide, the dominant force within astrobiology has, understandably, been the US, with its advanced space research program. However, attempting to compete with the US space program might be not only impractical but also unnecessary. Impractical, because of the level of funding that US research secures. Unnecessary, because localized competition to achieve various shared goals may be useful, but the overall relationship between astrobiology and Europe and in the US might better be thought of as a dialogue between significantly different interlocutors. Within this dialogue, the practice of astrobiology across Europe brings something distinctive to the discussion, i.e. a greater level of integration between scientific inquiry and other forms of inquiry which draw upon the social sciences, the arts and the humanities. This promises to enrich the discussion and our shared international understanding of the goals of astrobiology research.

The difference is, of course, one of degree. There is a classic image of the relationship between science and the humanities (in particular) as a troubled relationship. This is an image which draws upon a contrast of 'two cultures' set out at the very start of the US and Russian space programs, in C. P. Snow's influential 1959 Rede Lecture and then in his classic *The Two Cultures* (1963). Snow argued that the whole of the intellectual life of western society was split between the very different cultures of the sciences and the humanities. It was not something that he celebrated, but rather something that he considered a source of regret and also of various kinds of dangers. His sympathies lay more with one side of the divide than the other, and even more with an overcoming of the divide. On the one hand, it risked promoting technocratic solutions to social problems. On the other, it encouraged what Snow called the 'natural luddite', full of reservations and even suspicions about the impact of science. While Snow's instructive paradigm has remained influential, its limits have long been appreciated. In retrospect, it may seem like a very 1950s view of the world. A timely warning about a particular era, marked by advancing technology, and by a fear of technology and what it might bring (Fig. 1.1).

Longer-term trends, which the idea of two cultures does not entirely capture, have included pressures towards interdisciplinarity and multidisciplinarity, and towards the partial integration of science with other areas of inquiry. On the science side, it is difficult to work on the origins, evolution and future of life without drawing upon our best understandings of society and of all that the most impressive work in the arts and humanities brings. Science does not, after all, set its own framework for long-term impact. How it influences matters, and feeds into our lives, depends upon a broader set of considerations, sensitivity to which may enhance scientific practice and its social benefits.

This move beyond anything resembling a *strict* separation into two cultures, has been a marked trend in recent years, exemplified in work on space, with NASA and

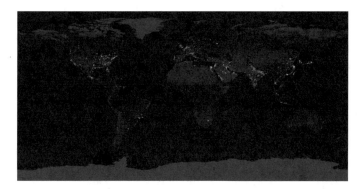

Fig. 1.1 Signs of human technology seen from outer space. The nighttime view of Earth. *Copyright* NASA Earth Observatory image by Robert Simmon. Retrieved from: https://goo.gl/SgNGkS

the Templeton Foundation jointly funding a project on the societal implications of astrobiology at the Princeton Center of Theological Inquiry between 2015 and 2017. NASA has also jointly-funded a Bloomberg Chair in astrobiology at the Library of Congress, a position inaugurated in 2012 and held successively in its initial years by a planetary scientist, three historians of science and an astronomer. A mix of figures from both sides of Snow's great cultural wall. The project of the astronomer and fifth chair, Lucianne Walkowicz, known for her work on stellar magnetic activity, covered "Fear of a Green Planet: Inclusive Systems of Thought for Human Exploration of Mars", with a strong emphasis upon inclusion, upon bringing together cutting-edge science and the lessons that we can learn from the many different kinds of histories of exploration of our own planet. What is addressed here is, of course 'histories' rather than 'history', it accommodates reasonable disagreement and the kind of dialogue that remains open to further insights.

This is integrated work at a high level, far beyond the more speculative attempts to combine science and the humanities that were available in earlier decades. It is an exemplar of the ways in which, since the beginning of our current century, work in astrobiology and its social implications have made massive strides forward, both in terms of the volume of work and, more importantly, its quality. This is not to dismiss earlier and more speculative efforts, but simply to point out the advantages that an integrated and established multidisciplinary field can offer, when compared with the insights of even the most innovative but isolated scholar.

Walkowicz's work is also an exemplar of greater cohesion in the conversation between the hard sciences (on the one hand) and the arts, humanities and social sciences (on the other). While there remains an understandable, perhaps even indispensable, strong focus in the US and elsewhere upon hard space science and technology, any division into two rigidly-separated cultures now seems artificial. Even the idea of 'hard' science as opposed to social science, is only useful up to a point. The addition of a third culture to the divided model, a culture of scientific outreach, where scientists directly address the public without the intermediary of the arts and humanities, does not obviously look more accurate than the original. Scientific outreach does not seem to work in that way but is, again, more inclusive and ordinarily seeks to draw from disciplines in the arts and the humanities. There are, of course, multiple contrasts that might still be drawn between the hard sciences and what sits outside of them, but also a greater sense that they belong to the same larger world of human enquiry.

The examples above are, of course, from the US. And indicate that we do not need to look to Europe in order to see the limitations of the Snow paradigm (A paradigm which, it may be re-emphasized, performed useful work in its day.). However, examples of this sort may also suggest that we should be cautious about replacing Snow's division between the two cultures of science and humanities with an equally problematic division between the scientific cultures of the US and of Europe. The White Paper that follows is the result of international co-operation across the latter, but involved colleagues from the US, from bodies such as NASA and the National Institute of Aerospace as well as from various US universities such as Clemson. One of the co-editors was, in fact, involved in the NASA-funded Princeton project mentioned above. Most of the contributors have collaborated with colleagues from the US on multiple projects and publications, with all of the lines of mutual-influence this helps to create. Indeed, editors of volumes and special editions on matters related to space and to astrobiology, will typically make sure that their geographical spread of contributors is good in order to reflect the broader, global state of the field. Astrobiology is an inherently international discipline and brings together scholars on a global scale. The pressures towards an inclusive approach are present internationally. They are not the preserve of European researchers.

However, the degree of integration between work on astrobiology and society has nonetheless been, in some important respects, a distinctive feature of the European research community. Indeed, while there have been a number of collaboratively produced overviews of the emerging field of astrobiology in both Europe and the US, this White Paper is the first comprehensive overview of the *societal implications* of astrobiology in either. It does not seem accidental that it has been produced in Europe rather than anywhere else.

The White Paper itself is a collaborative product which spans disciplinary divisions, with scholars drawn from twenty-eight countries and from departments of Philosophy, Sociology and Theology (among others), as well as from hard sciences departments, the European Space Agency and the National Centre for Space Studies in France. It is particularly notable that several of the key contributors to the emerging field of the ethics of space (from both Europe and the US) have been involved in the text, and have participated in the multiple events across Europe which have fed into it. The associated series of conferences, workshops, and short-term scientific missions (STSMs) funded by the EU program Cooperation in Science and Technology (COST) as part of the Cost Action TD1308: Origins and evolution of life on Earth and in the Universe (2014 to 2018) and its Working Group 5: The history and philosophy of astrobiology, has been pivotal to the development of astrobiology in Europe as a strongly inclusive and integrated discipline drawing upon the highest levels of scholarly work. These events have also been pivotal to the case for establishing a European Astrobiology Institute (EAI) in order to further the collaborative efforts of the international research community within the European Research Area and beyond. The White Paper serves as one of the founding documents of this EAI, and aims to clarify the initial scope of its work. It is not simply a plea for cohesion and research integration, it is an exemplar of both, an example of what is possible.

The key areas finally settled upon, and grouped together as chapters, are not intended to exclude all other matters. Rather, they form a core of research topics of particular strategic significance (for the coordination and dissemination of astrobiology within Europe). While this introduction has focused upon the US/Europe contrast, in order to help sharpen the sense of a distinctive European contribution to the discussion, the topic areas in the White Paper address the much broader issue of humanity's place within the development of life. Chapter 2 introduces astrobiology to the reader and focuses upon the European context and the ways in which the humanities and social sciences and hard science combine. Chapter 3 deals with the broader international context of work in the US, Asia, Africa and Latin America, and introduces the shared regulatory framework for planetary protection. Chapter 4 deals with the combined topics of how we think about humanity's place in the universe (inclusive of religious perspectives) and the challenges of disseminating and popularizing work on astrobiology within society at large. (This covers the issues of outreach.) Chapter 5 deals with the problems of situating issues such as 'planetary protection' in its restricted legal sense, in relation to environmental discourse and to the idea of sustainability. The chapter does not seek to resolve the challenges of working with narrower and broader understandings of planetary protection, but merely seeks to clarify some of the issues involved. Chapter 6 deals with the vital practicalities of education, training and the cultivation of scholarship, matters crucial to the sustainability of progress in the research field of astrobiology itself. Chapter 7 introduces some of the key stakeholders in the commercial and technological sectors and addresses the possibility of conflicts of interest (where the interests in question may all be legitimate but not always identical). Chapter 8 integrates astrobiology research and the study of astrobiology itself, as a science. Chapter 9 ends the main body of the text by addressing the challenges facing the European astrobiology community, and the ways in which a European Astrobiology Institute can help the research community to meet them. The text closes with an Afterword Chapter 10, addressing the timeliness of the proposal and the connection between scientific (and broader social) interest in life (its origins, evolution, and distribution) and the wider development of human activities in space, with European initiatives placed at the heart of such activities. The overall intention is, therefore, to situate the European research in a distinctive way, but within the broader global context of work within the field, within the context of an expanding range of international activities in space, and within scholarly work on our understanding of humanity.

There are many valid ways of putting together a project of this scale and ambition. One way would be to set out a series of personal visions for what astrobiology is and what it may become. Such an approach could encourage illuminating speculation and might, in a modified form, lead to various lines of new research. The White Paper takes a different approach. It does not represent the views of any single person but is, rather, a consensus document about the societal implications of astrobiology and about various social challenges that astrobiological research is likely to face in the future. Challenges such as thinking about astrobiology in the context of planetary protection and environmental protection, and making provision for effective outreach in an age of social media where science and theories that merely mimic the practice of science may not always be easy to separate. In the interests of cohesion, focus and outcomes, the contributors have been encouraged to work to a shared underlying three-part structure of background, challenges and suggestions. This approach is geared not only to the formulation of new research proposals, but also addresses the institutional needs of the astrobiology community and policy formation.

Chapter 2
Astrobiology and Society in Europe

**D. Dunér, K. A. Capova, M. Gargaud, W. Geppert, A. Kereszturi
and E. Persson**

2.1 Introduction

There is only, as we know it, one planet with life—our own Earth. However, current research in astrobiology searches for a second sample of a living world. Astrobiology, which concerns the origin, evolution, and future of life here on Earth and beyond, has become a rapidly expanding research field during the last two decades. European researchers are playing a leading role. Thousands of planets in other solar systems have been discovered. Knowledge about life's evolutionary origin, and its requirements and environmental conditions have expanded considerably. It is not unlikely that one day—some say that this could happen within the next few decades—we will discover evidence of the existence of another living planet. Living or fossilized microbes could be found within our Solar System, or we could find signs of biological processes on planets in other solar systems. But even if this never happens,

D. Dunér (✉)
History of Science and Ideas, Lund University, Lund, Sweden
e-mail: David.Duner@kultur.lu.se

K. A. Capova
Durham University, Durham, UK

M. Gargaud
Laboratoire d'Astrophysique de Bordeaux, Pessac, France

W. Geppert
Physics Department (Fysikum), Stockholm University (Stockholms universitet), Stockholm,
Sweden

A. Kereszturi
Csillagászati ès Földtudományi Kutatóközpont (Research Centre for Astronomy and Earth
Sciences), Budapest, Hungary

E. Persson
Department of Philosophy, Lund University, Malmö, Sweden

© The Author(s), under exclusive license to Springer International Publishing AG,
part of Springer Nature 2018
K. A. Capova et al. (eds.), *Astrobiology and Society in Europe Today*, SpringerBriefs
in Astronomy, https://doi.org/10.1007/978-3-319-96265-8_2

7

Fig. 2.1 Flammarion Engraving. A historical interpretation of man's quest of knowledge and understanding of the universe. *Copyright* Public Domain. Camille Flammarion: *L'Atmosphère: Météorologie Populaire*. Paris, 1888, p. 163. Retrieved from: goo.gl/E9gF8U

astrobiological research will still give us a new understanding of how life emerged on our planet, how it evolved, and what environmental conditions it needs in order to survive. In all, current and future research in astrobiology will change the view of how humans look at themselves, what it means to be a human, to be alive, to survive, where we come from, and where we are heading (Fig. 2.1). Astrobiology has clear existential implications, but beyond these, it also has concrete cultural, ethical, societal, educational, political, economic, and legal consequences. How will the general public react if we discover life on another planet? What pedagogic role can astrobiology play in elementary and higher education? To what extent should we utilise space for commercial and industrial purposes? How should this be politically managed and how should it be legally regulated? This White Paper on the societal implications of astrobiology research in Europe, which is a joint interdisciplinary effort of Working Group 5 within the COST Action TD1308 "Origins and Evolution of Life on Earth and in the Universe", aims to gather together these challenges and implications, and in so doing lay the ground for a European Astrobiology Institute.

2.2   What Is Astrobiology?

Astrobiology, the study of the origin, evolution, distribution, and future of life, is a multidisciplinary activity that involves natural sciences such as biology, astronomy, chemistry and geology. Astrobiology encompasses the search for life elsewhere in the universe and in terrestrial environments which are analogous to those of outer space. The study of extreme life forms, extremophiles, is of great importance to astrobiologists in these efforts to expand our view of the potential for life elsewhere. The scope of astrobiology research is framed by the following set of questions:

- Where, when, and how did life emerge and evolve on Earth?
- What are the conditions under which life can exist?
- Does life exist elsewhere and, if it does, how can it be detected?
- What is the future of life in the universe?

2.3 Organising EU Astrobiology Research

Astrobiology has the potential to be a flagship of European cooperation in science. It provides an ideal ground for collaborative European projects which support the ethos of cooperating countries. Making astrobiology a visible part of European science should therefore be a priority for the EU.

The establishment of a European Astrobiology Institute will provide the perfect forum for such collaborative efforts and should be a key priority for European research institutions as well as the European astrobiology community. To have an active astrobiology research programme, coordinated and fostered by such an institute, will enhance the international standards of European space research, and of European science in general.

This White Paper recognises the establishment of the EAI as such a priority and is designed to be one of its founding documents. It also acknowledges the societal implications of astrobiology, and thus the role of the social sciences and humanities in optimizing the positive contribution that astrobiology can make to the lives of the people of Europe and the challenges they face.

A dedicated European Astrobiology Institute should coordinate astrobiology activities across Europe and benefit from interdisciplinary and multidisciplinary engagement. The EAI will be well placed to promote astrobiology research, assist in the decision-making process of relevant European institutions, be involved in mission planning, engage in science dissemination, education and communication, as well as outreach and media work. The following would be appropriate tasks:

- Perform cooperative cutting edge research in Europe, focussing on a set of agreed central themes.
- Coordinate astrobiology research and related scientific activities in Europe.
- Support interdisciplinary networking and interactions between researchers from different fields.
- Coordinate European training, education and outreach activities in astrobiology.
- Act as an external advisor for ESA and other European research bodies.

The EAI will thus function as a distinct entity to coordinate, promote and further develop EU research in astrobiology. It will cover all aspects of astrobiology research on a European level and strive to sustain the European Research Area's role as a world leader in this comparatively new but established and rapidly developing field.

2.4 The Contribution of Humanities and Social Sciences to EU Astrobiology

The aim of this document is to lay the ground for a coordinated multidisciplinary European Astrobiology Institute (EAI) and to establish the contribution of the social sciences and humanities to astrobiology research in Europe. It is envisioned that the humanities and social sciences will make the following contributions to EU Astrobiology:

- Assist in the shaping and modification of foundational scientific concepts such as 'life', and provide useful social and historical frameworks for these concepts.
- Support the foundation of the European Astrobiology Institute and assist in shaping the future of European astrobiology research.
- Anticipate and prepare for novel ethical challenges related to the conduct of science; and assist in making informed decisions on astrobiology research.
- Provide insights into popular understandings of astrobiology and support astrobiology communication, dissemination, education and training.
- Initiate and conduct relevant social science and humanities research actives in a European context; conduct research into societal perceptions of astrobiology and into the immediate effects of the advancement of science on society.
- Explore the existential implications of scientific findings.
- Disseminate research results to the public and produce materials for dissemination; increase astrobiology awareness and literacy.
- Take part in educational activities such as summer schools, outreach events, and engagement with mass media.
- Assist in the production of educational materials as well as school curriculum development, training programmes and teacher training.

Chapter 3
The International Context
of Astrobiology

E. Persson, A. Anglés, L. Billings, E. Nabulya, S. Ramos, K. Smith
and S. Tirard

3.1 Introduction

Astrobiology is an internationally established field with research taking place on all continents, including Antarctica, by research teams from all over the world. The International Astronomical Union (IAU) established a commission for what was then called 'Bioastronomy' in 1982. Its successor, the Astrobiology Commission of the IAU was established in 2015.

The USA is currently the main leading nation within astrobiology with the largest number of researchers as well as the highest level of research output. This is at least partly an effect of the establishment of the NASA Astrobiology Institute (NAI) in 1998 to promote cooperation and provide meeting grounds and funding for the research community. European research teams are, however, already playing an important role and have provided several major breakthroughs in the field. In order

E. Persson (✉)
Department of Philosophy, Lund University, Lund, Sweden
e-mail: erik.persson@fil.lu.se

A. Anglés
University of Hong Kong, Pokfulam, Hong Kong

L. Billings
NASA's Astrobiology Program, California, USA

E. Nabulya
Makerere University, Kampala, Uganda

S. Ramos
Universidad Nacional Autónoma de México, Mexico, USA

K. Smith
Clemson University, Clemson, USA

S. Tirard
Université de Nantes, Nantes, France

© The Author(s), under exclusive license to Springer International Publishing AG, part of Springer Nature 2018
K. A. Capova et al. (eds.), *Astrobiology and Society in Europe Today*, SpringerBriefs in Astronomy, https://doi.org/10.1007/978-3-319-96265-8_3

11

for the European research community to become an even more highly regarded actor and valued cooperation partner in the field, we believe that it is important to draw lessons from other countries and adapt them to a European setting. The establishment of the NAI was an important research infrastructure initiative. However, in order to work in a European setting, and draw from our areas of strength, the idea of an institute needs to be adapted. Europe has essential and unique strengths in astrobiology: a network of national and international scientific institutions, with high-level skills proven by several space missions and programs; a community of lawyers and a distinctive legal tradition involved in the elaboration and evolution of space law; an extensive academic network, already committed to reflection on the scientific and social dimensions of astrobiology.

An important strength is the European tradition of bridging the 'two cultures' of science and the humanities, and integrating the practice of science with social responsibility. Although there are crossovers between the two in other parts of the world, such a distinctive ethos of interdisciplinarity and multidisciplinarity is less visible elsewhere. Below, it is recognised as a limitation of the, otherwise strong, US experience.

3.2 European Astrobiology Research to Date

The European Space Agency (ESA) was founded in 1975 and its missions and projects have helped to promote the emergence of dynamic research on space in Europe. The International Space University (ISU) was founded in 1987 in Strasbourg, dedicated to the research and development of outer space exploration for peaceful purposes, through international and multidisciplinary education and research programs. Astrobiology has its place in most ESA programs, and is therefore supported by substantial financial and technological investments from its member states. The Rosetta mission's confirmation, in 2016, of a connection between comets and the life-supporting atmosphere of the Earth illustrates the point. So too does the ongoing ExoMars Program, a joint astrobiology project of ESA and the Russian space agency Roscosmos with a focus upon biosignatures and habitability (Fig. 3.1).

Fig. 3.1 North to south. Image of Mars taken by ESA's Mars Express during camera calibration. *Copyright* ESA/DLR/FU Berlin, CCBY-SA3.0IGO. Retrieved from: https://goo.g l/ccrKJv

To help coordinate work in the field, the European Astrobiology Network (EANA) was created in 2001 and has organised a yearly European workshop on astrobiology as well as an online training course. Other permanent networks involved in astrobiology have emerged. For example, the Astrobiology Society of Britain (ASB) was created in 2003; the Italian Astrobiology Society (SIA) in 2006; the Nordic Network of Astrobiology (NNA) was established in 2007; the French Astrobiology Society (SFE) was created in 2009, and the German Astrobiology Society (DAbG) in 2016.

Since 2014 the COST Action "Origins and evolution of life on Earth and in the Universe" (ORIGINS) has brought together scientists from more than 20 European countries. At several European research institutions (e.g. the universities of Edinburgh, Szczecin, Stockholm) astrobiology centres have been launched. Furthermore, an association of students and early career scientists (AbgradE) has been founded. These networks help to organize conferences, grant bids, summer schools and outreach events. They are characterized by interdisciplinarity. As a result, the community of European scientists interested in astrobiology is building a European identity for work in this field.

A number of new initiatives have also recently emerged. The COST Action ORIGINS has been a welcome boost to the European astrobiology community, organizing a series of well attended meetings and providing funding for research visits by European scientists and students as well as training events. Also, the Action has funded short-term scientific missions for scientists and held yearly conferences and specialised workshops. The different working groups of the Action have been active in discussing and pursuing new research avenues and organizing focused workshops. For example, the working group on historical, ethical and philosophical questions in astrobiology has been instrumental in bringing researchers from science and the humanities together through the organization of a workshop and an associated summer school in Southern Sweden and through follow-up engagement. Last, but not least, this book is one of the outcomes of this COST Action.

The European Astrobiology Campus, launched as a Strategic Partnership under the Erasmus + programme has provided a coherent and comprehensive training programme in astrobiology. One of its main activities has been to continue and extend the successful series of summer schools organized by the Nordic Network of Astrobiology and the French Society for Exobiology. These events have offered training for future European research leaders in astrobiology, by world renowned scientists, at astrobiologically interesting sites in the field. They have also enabled the attendees to find future cooperation partners and to start their own research projects based on the practical exercises carried out in the summer schools. In a multidisciplinary field it is also important to train the trainers and to foster the development and exchange of innovative teaching methods. To this end, the European Astrobiology Campus and the IAU Commission on Education co-organised a conference on Education in Astronomy and Astrobiology in Utrecht in July 2017. Furthermore, workshops on outreach have been organised and the leading reference work, the Encyclopedia of Astrobiology, was co-sponsored by the initiative. The funding for the COST Action and the European Astrobiology Campus is now at an end, officially ending in spring 2018 and autumn 2017, respectively.

These initiatives demonstrate the need for greater overall structure and coordination for astrobiology research in Europe. More specifically, the research dynamic is characterized by:

- The need for a more structured interdisciplinary network covering a large spectrum of scientific disciplines and including both the humanities and social sciences (e.g. sociology, anthropology, law, political science, theology, history, and philosophy).
- The need for a sustained coordination of international projects.

These developments and challenges help to provide reasons for establishing a European Astrobiology Institute. The institute will emerge out of the current networks and will be the best tool to achieve the more ambitious goals of European-wide scientific cooperation. Whereas the EAI will be a consortium of institutions, EANA is (and will remain) a society of individuals. The president of EANA and the four other EANA Executive Council members are also members of the Interim board of the EAI. The discussions on the establishment of the EAI have been held at several EANA workshops.

As a stronger involvement of astrobiology is expected within science in general, societal impact should also be considered seriously, in its own right, and as part of the ongoing dialogue about social responsibility and the broader impact of science and technology. The Institute will have to deal with these issues.

3.3 American Experience

While NASA cannot provide the kind of support necessary to foster truly systematic investigation of the societal implications of astrobiology, it has sponsored some important US projects along these lines. However, they have been sparse, sporadic, and disconnected. The first formal investigation of these kinds of questions was a NASA-sponsored symposium held in 1972, which focused on "the social, philosophic, and humanistic impact" of the discovery of extraterrestrial life. However, the US did not capitalize on this promising beginning, focusing instead on the Shuttle and International Space Station programs. Only in 1996, following claims of fossil evidence of Martian life in a meteorite, did growing scientific, political, and public interest prompt NASA to attend to the societal implications of astrobiology.

NASA's Ames Research Center held a workshop on the "Societal Implications of Astrobiology" in 1999. In 2003-04, NASA's Astrobiology Program co-sponsored (with the American Association for the Advancement of Science (AAAS) Dialogue on Science, Ethics and Religion (DOSER) program) a series of workshops to address the "philosophical, ethical, and theological implications of astrobiology". Then in 2009, the Search for Extraterrestrial Intelligence (SETI) Institute in California held a workshop to develop a roadmap of societal issues relating to astrobiology. NASA's Astrobiology Institute approved the formation of a focus group on the social implications of astrobiology in 2012. However, this group foundered in 2015 and there are no plans to revive it.

Despite these challenges, in just the last few years, the NASA Astrobiology Institute (NAI) has undertaken three major new initiatives designed to broaden the community of scholars currently working on these issues. In collaboration with the Kluge Center of the US Library of Congress, NASA Astrobiology Institute is funding the Baruch S. Blumberg NASA/Library of Congress Chair in Astrobiology. The Blumberg Chair makes it possible for a senior researcher to be in residence at the Kluge Center, to make use of the Library of Congress collections, and to convene programs that ensure the subject of astrobiology's role in culture and society receives considered treatment. The Chair has resulted in a series of congressional hearings on astrobiology held by the House Science, Technology, and Space Committee as well as two books, three interdisciplinary dialogues, and several public symposia.

From 2015–17, NASA also provided support for an exploration of the societal implications of astrobiology being conducted by the Center of Theological Inquiry (CTI) in Princeton, New Jersey. For this project, CTI gathered together a small group of researchers in theology and other disciplines, who spent an entire year in residence working on these kinds of questions. Finally, the 2015–16 NASA Astrobiology Debates was a year-long academic project for university and secondary education students involving in-person and online debate tournaments, speech competitions, public exhibition debates, topic-expert panels for student audiences, and student-conducted topic interviews with a cross-disciplinary group of subject-matter experts. The debate topic for this project was: "Resolved: An overriding ethical obligation to protect and preserve extraterrestrial microbial life and ecosystems should be incorporated into international law." The aim of the debates project was to stimulate student, teacher, and school research and dialogue on astrobiology in preparation for these events and at the events themselves.

Despite these promising recent developments, the number of researchers in the social sciences and humanities who have engaged in these activities is still relatively small. What is more, since NASA can make no long term commitment to funding this kind of activity and research, interested investigators are uncertain what, if any, support will be available in the future.

3.4 International Outlook

3.4.1 Africa

Astrobiology has not, to date, been a major research area in Africa. There are, however, local exceptions. One is the Palaeontological Society of Southern Africa which spearheads research on fossils, for instance the Barberton fossils in the Mpumalanga region. This allows them to consider ideas on the origin of life. Another is the Ibn Battuta Centre for Exploration and Field Activity in Marrakech, Morocco, which is operated by the European research network Europlanet and is used mainly as an analogue environment for testing equipment and strategies for Mars missions. 'Astrobiology' is mentioned on the website for the Working Group on Space Sciences in Africa but without any details, and it is not mentioned in the Mombasa Declaration on Space and Africa's Development, an important statement on the future of research in Eastern Africa. Nor is it among the areas of interest listed by the Astronomical Society of Southern Africa (ASSA). Notwithstanding, there is a detectable interest throughout Africa in issues concerning extraterrestrial life in cultural products such as films, fiction and fine art.

3.4.2   Asia

China and Japan are both active in astrobiology research. Both are fairly well represented at international astrobiology conferences, and both are involved in international cooperation with European research groups and individual European researchers. Important international astronomy conferences have been held in Japan, Vietnam and India, and the International Conference on Astronomy and Astrobiology in 2018 takes place in Osaka, Japan. The Astronomical Research Institute of Thailand (NARIT) is cooperating with its Asian partners, such as the Astrobiology Center (Japan), Beijing Planetarium (mainland China) and KASI (South Korea) to organise outreach events related to astronomy and astrobiology for high school and university students, teachers and the general public. The Thai-Asian Astronomy Forum also receives support from other organizations to organise talks and events about Astrobiology and Life in the Universe. The Earth Life Science Institute (ELSI) at the Tokyo Institute of Technology is already one of the most active research institutes in the field and continues to invest substantially in active research, research facilities and in recruiting top researchers from Japan as well as Europe and elsewhere. Since 2015, the Japanese Space Agency (JAXA) and a number of Japanese universities have run an experiment at the International Space Station (ISS) investigating the potential for interplanetary transfer of organic material. One of the missions for China's new Five Hundred Meter Aperture Spherical Telescope (FAST) will be to search for radio signals from Extraterrestrial Intelligent Life, partly backed by the Breakthrough Initiative (a body involving Mark Zuckerberg and the late Stephen Hawking). There is also a newly initiated cooperation between Europe and China, the ISSI Astrobiology Team, aiming at accelerating the growth of astrobiology in China. The University of Hong Kong, one of the members of the International Space Science Institute (ISSI) Astrobiology Team, is a top-level research institution where researchers are focused in the detection of life beyond Earth and the origin of life on Earth. India and South Korea also have active research groups and research centres as well as some ongoing educational work in astrobiology. In 2016, the International Centre for Interdisciplinary Science Education in Vietnam organised an international astrobiology conference in cooperation with European researchers.

3.4.3   Latin America

Astrobiology research in Brazil dates back to 1958 with Brazilian astrobiology focusing on extremophiles and biosignatures on exoplanets. NAP-Astrobio is an online discussion forum based at the University of São Paulo, covering these and related topics, spanning extremophiles, extraterrestrial environment modeling, and habitability. The Mexican Society of Astrobiology (SOMA) is also pursuing multiple lines of astrobiology research and organises the Mexican School of Astrobiology. The majority of Mexican astrobiology researchers are associated with the National Autonomous University of Mexico (UNAM). Researchers at the National University of Colombia are also addressing astrobiological issues from both scientific and philosophical perspectives.

Chile is the host country for the European Southern Observatory, with the latter playing an important role in exoplanet research. Researchers in Atacama Biotech Astrobiology and Biotechnology are working on projects in the Atacama Desert, which is considered a good analog model for Mars. As a result of its profile in the field, Chile was also the host country for the 2017 IAU astrobiology conference. Peru has several astrobiology oriented organisations and has hosted meetings on astrobiological issues. The La Joya desert, located in the Arequipa region of Peru, is a Mars analogue. The University of Lima gives classes and workshops discussing links between astrobiology and philosophy. The Scientific Society in Peru is also aiming to develop an Institute of Astrobiology in Latin America. Argentina is already home to two research groups in astrobiology with both groups working on simulation of interplanetary conditions. They have developed a device for use with future space missions that can probe for evidence of life in other parts of the Solar System.

Chapter 4
Society, Worldview and Outreach

K. A. Capova, L. Dartnell, D. Dunér, A. Melin and P. T. Mitrikeski

4.1 Introduction

As well as impinging upon issues of law and governance, astrobiology is also bound up with questions concerning who we are and where we come from, worldview questions of a more existential and philosophical sort. The questions that it seeks to tackle have, for centuries, been central to the humanities and to social science disciplines.

The social significance of astrobiology stands out in its impact on forms of social life such as religious beliefs, spiritual commitments and the worldviews of contemporary Europeans. The question 'where do we come from?' is a common theme in theoretical cosmogonies of origin that explain the creation of the world (creation myths) and origin of the humankind globally. They are a central component of human self-understanding and a cultural frame of reference in worldview formation.

Given the social and cultural significance of these questions, their public appeal and attractiveness to mass media, astrobiology carries the potential for academic and

K. A. Capova (✉)
Department of Anthropology, Durham University, Dawson Building, South Road, Durham DH13LE, UK
e-mail: k.a.capova@durham.ac.uk

L. Dartnell
University of Westminster, London, UK

D. Dunér
Lunds universitet, Lund, Sweden

A. Melin
Malmö Universitet, Malmö, Sweden

P. T. Mitrikeski
Institut za istraživanje i razvoj održivih ekosustava (Institute for Research and Development of Sustainable Ecosystems), Zagreb, Croatia

© The Author(s), under exclusive license to Springer International Publishing AG, part of Springer Nature 2018

K. A. Capova et al. (eds.), *Astrobiology and Society in Europe Today*, SpringerBriefs in Astronomy, https://doi.org/10.1007/978-3-319-96265-8_4

19

Fig. 4.1 An exoplanet or extrasolar planet. An artist's impression of the Jupiter-size extrasolar planet, HD 189733b, being eclipsed by its parent star. *Copyright* ESA, NASA, M. Kornmesser (ESA/Hubble) and STScI. Retrieved from: https://goo.g l/Kxm5UB

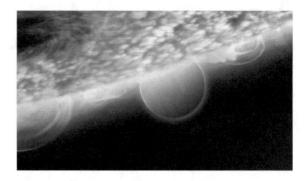

social impact. However, the far broader social implications of these questions and the role they play in contemporary debates, are reflected in numerous literary works, in films, plays, gaming industry, music, and fine arts. They are integral to contemporary (Western) society and popular culture.

4.2 The Public Understanding of Astrobiology

The role of the social sciences and the humanities in relation to astrobiological research and its public reception can be divided into four broad areas: clarifying key scientific concepts and providing them with background narratives; understanding, and anticipating, the social impact of science; astrobiology outreach and popularisation; and science education and training. Astrobiology as a science and as the production of new knowledge may change the ways in which we situate ourselves in the world, but it is likely to do so at least in part through rapidly evolving communication channels and digital media. As examples, the worldwide reporting of discoveries of liquid water on Mars, and of new extrasolar planets, have occurred in real time and have been subject to all of the kinds of scrutiny, modification and reinterpretation that our developing technologies allow (Fig. 4.1).

Astrobiology as a newly formed multidisciplinary field of scientific enquiry plays an important role in the study of knowledge production and science as a practice. The history and philosophy of science contribute to an understanding of the dynamic processes of science development. New fields of social research, such as Science, Technology and Society (STS) and the anthropology of science, explore how society (on the one hand) and science, research, and technology (on the other) affect one another.

With the power of social media on the rise, it can provide what may seem almost the opposite to the academic peer review process. Social media make science easily distributable yet they may also affect its credibility and the reliability of the information that is shared, modified and often reproduced in altered contexts. Making sure that information is distributed in a reliable way to the interested public, as well as to classrooms, will be a challenge. There is a need for scientists, as well as decision makers and planners, to understand how the public responds to science and how it understands new discoveries in basic research. There is also a need to address at least some popular beliefs that mimic certain aspects of science, especially beliefs that tend to compromise the practice of actual science and the reliable dissemination of its results.

The social sciences and humanities play a key role in understanding, anticipating, and handling misconceptions about scientific research. Their more qualitative approach may provide understandings of social processes and help to provide insights into how the workings of astrobiology may be communicated to a broader audience than that of practicing scientists.

4.3   Popularisation, Dissemination and Outreach

Science engagement with the general public is a crucial component of academic research. Much research is publically funded, and thus ultimately paid for by taxpayers. Outreach and dissemination are part of the reasonable transparency and openness of science that the public are financing. A public that is engaged with science may also be better placed to appreciate it, better placed to recognise the value of conducting diverse research, and better placed to comment on the practice of science from an informed viewpoint. The importance of this has been highlighted in recent years through unwarranted but widespread suspicion of genetic engineering and nanotechnologies.

Outreach can provide school students with captivating events and role models to inspire them to continue Science, Technology, Engineering, Mathematics (STEM) subjects through to secondary education and university. This is vital for the information and technology economy as well as for training the next generation of research scientists. The potential for astrobiology outreach will differ from that for medical or healthcare research, and may have a different appeal. But this diverse interdisciplinary research field is intrinsically interesting to the general public and to school students. It also has a large potential for attracting positive attention to science.

Given the appeal of speculation on our origins, and the search for life beyond Earth, astrobiology has a large potential to contribute to producing not only a scientifically skilled workforce but also a scientifically literate public, with the further benefit of orienting young people toward technology-related domains alongside science.

However, this interest in the subject-matter of astrobiology means that the field is also vulnerable to misinterpretation, or even mistaken association with pseudo-scientific theories. A special challenge is presented by mass media which tends to favour fascinating stories about life elsewhere. A common misconception is the conflation of astrobiology and the search for life with UFOs and abduction stories. Fascination and interestingness can take the place of concern for reliable evidence and truth, with basic levels of scientific accuracy being compromised.

A further challenge for astrobiology communication concerns the response to an actual discovery of clear evidence for life beyond our planet. Answers to key questions concerning the responsibilities of the media, the scientific community and government authorities are unclear. Particularly so over key questions of how best to respect a public entitlement to know, how to inform the general public about a successful detection of extraterrestrial life, and when to do so. Press conferences before peer review have often turned out to be premature and can harm confidence in astrobiology and commitment to the funding of viable forms of research. A delay in making a discovery of life public, until the result has been through the peer review process, could on the other hand result in rumours and feed conspiracy theories. A set of guidelines or protocols stating how to proceed could prove useful.

Another challenge, but also an opportunity, is to engage the citizen science community even more in the process of searching for habitable planets and to further engage the public in the co-creation and open use of research data by exoplanetary surveys. Commitment to doing so is a core value of the Open Science, Open data and Open to the World policy, as stated by the European Commission.

The internet is a very powerful tool for research and for the dissemination of results. It has also increased the rapidity with which views can spread in a 'viral' or 'meme-like' way. It is, therefore, important for astrobiology researchers to engage with the public through such media in order to present the discipline as it truly is, the scientific methods that are used, and the genuinely exciting discoveries being made. To effectively manage situations of major discovery and instances of viral misreporting would require, at least, a statement by a reputable astrobiology community body in response to the dissemination of any inaccurate information, especially where the latter carries the potential to affect future funding decisions and the conduct of science. Overall, a unified strategy for astrobiology communication is needed alongside active engagement in astrobiology outreach and the dissemination of research results at the international level. Such a strategy would involve, firstly, the development of a dissemination plan to raise awareness of the EAI and the work of the astrobiology community within the EU in, and among, target audiences and stakeholders. This would help to maximise the impact of communication efforts. Secondly, such a communication strategy would raise awareness of work on astrobiology within the broader science community and help to develop models for outreach and the dissemination of astrobiology to schools, the general public, and private companies, using available communication channels known to be favoured by the relevant audiences.

4.4   Rethinking Humanity's Place in the Universe

Astrobiology can help to answer longstanding questions about whether or not we are alone in the universe and about where we have come from, where we are going, and how life first emerged. As indicated above, it responds to questions of a more philosophical and existential sort. We might even say that, although it is a scientific discipline, it responds to the deep longings of humans as social beings. The separation of disciplines that is an unavoidable feature of a classic system of education, specialization and academic training, leads many of us to set aside questions of this kind at some point. These fundamental questions can, however, keep people motivated in searching for answers, and keep people open minded and oriented towards scientific and technical fields that promise some kind of reliable answer as well as supporting economic development by innovation.

Astrobiology challenges our understanding of ourselves as human beings in the Universe. The search for life elsewhere touches on fundamental hopes and fears, on what it means to formulate a theory, grasp a concept, and develop our imagination. Contemporary astrobiology is, however, more than the result of discoveries and theories. It is the product of multiple and complex social and cultural processes. These include collaborations and technological changes which have made the current goals of astrobiology ambitious but realisable. The future success of astrobiology rests on how well the astrobiological community can perform efficient collaborations, build up supporting institutions, and integrate new technologies into research and into the spread of new knowledge. The challenges are socio-cultural as well as scientific.

As astrobiology is a human endeavour to answer profound existential questions about humanity, the consolidation and further success of astrobiological research would benefit from a recognition of the societal, and cultural factors involved in science. Among other things, it would benefit from closer consideration of the ethics of space exploration, questions about the nature of knowledge, and about the historical emergence of astrobiology as a scientific enterprise. Research in the humanities and research in the social sciences are therefore necessary contributions to astrobiology and complement the relevant ongoing work within biology, chemistry, geology, and astronomy.

4.5 Astrobiology and Religion

Religion is an important influence upon beliefs, behaviours and sense of belonging. It can influence people's perception of astrobiology and the prospects for a discovery of life elsewhere. The latter could have an unsettling effect on particular faith groups, although perhaps less so for Buddhists (who often believe in many worlds), and Muslims (given the apparent commitment to such life in the Qur'an). Based on an interpretation of the ancient Biblical poem, The Song of Deborah, several well-known Jewish rabbis have argued that intelligent extraterrestrials exist. According to Judaism, salvation is also open to all creatures with moral choice, and it is not dependent on the existence of one unique saviour. Some (not all) Christian denominations face a larger challenge.

For Christians, the possibility of microbial life outside of the earth raises the theological question of whether there might also be intelligent life elsewhere, and beings also in need of redemption. For evangelical Christians, committed to some notion of a divine plan, the existence of microbial life on its own, playing no further role, could be difficult to accommodate. However, the possibility of more advanced life elsewhere gives rise to the theological problem that there could also be more than one saviour. Historically, many mainstream theologians within Western Christianity have denied the existence of extraterrestrial life for precisely this reason: the uniqueness of Christ's redemptive mission seems to rule out the possibility of intelligent life forms on other planets. However, some Christian traditions in Europe, such as the Orthodox Christianity of the East, might place greater emphasis upon love and divine glory, and may more readily accommodate discovery since it can change little in the traditional beliefs of the Eastern Church. However, the split here may not be East/West, but rather along the lines of Protestant evangelical (where skepticism about life elsewhere has been most concentrated) and other forms of Christianity. Yet, even this is a simplification of complex patterns of religious attitudes, with many Protestant evangelicals perfectly open to the prospect of extraterrestrial life, and ready to accommodate its possible theological implications.

Research about religious views of extraterrestrial life (however microbial) has tended to be primarily historical and focused on canonical texts. Thus, there is a need to extend and incorporate more sociologically oriented empirical research about religious believers' views about such life. The results of such research should be communicated to religious bodies and to the public.

Chapter 5
Environment and Sustainability

E. Persson, J. Martínez-Frías, T. Milligan, J. Arnould and G. Kminek

5.1 Introduction

There are strong links between astrobiology and environmental concern. Astrobiology is the study of the origin, evolution and distribution of life in the universe—including Earth. Understanding life, and in particular the basic conditions for life, is important for our ability to create a sustainable future on Earth. The connection goes both ways, however. The preservation of biodiversity and of pristine environments on Earth is of the greatest importance for our ability to study life, its origin, distribution and future. Of special interest from an astrobiology perspective is the preservation of areas with conditions that can serve as analogues to extraterrestrial environments, areas with conditions similar to those under which life originated on Earth, and in general environments where extreme adaptations can be studied. Astrobiology also presents some direct environmental challenges that need to be considered, namely in the form of forward and back contamination. Both issues need to be approached from a technical perspective, but also from a societal perspective. And both must be understood within a broader context of ensuring the sustainability of practices, both scientific and commercial.

E. Persson (✉)
Department of Philosophy, Lund University, Lund, Sweden
e-mail: erik.persson@fil.lu.se

J. Martínez-Frías
Instituto de Geociencias, IGEO (CSIC-UCM), Madrid, Spain

T. Milligan
Department of Theology and Religious Studies, King's College London, London, United Kingdom

J. Arnould
Centre national d'études spatiales (National Centre for Space Studies), Paris, France

G. Kminek
European Space Agency, Noordwijk, The Netherlands

© The Author(s), under exclusive license to Springer International Publishing AG, part of Springer Nature 2018
K. A. Capova et al. (eds.), *Astrobiology and Society in Europe Today*, SpringerBriefs in Astronomy, https://doi.org/10.1007/978-3-319-96265-8_5

5.2   Environmental Protection and Sustainability

While the prospects for commercial mining in space may be good, and while space may seem to offer inexhaustible resources for scientific investigation and commercial use, the resources which are actually available at any given time are likely to remain limited and unevenly distributed. Without direct access to the asteroid belt, there are only a small number of metallic asteroids whose trajectories might allow them to be mined. Similarly, there are only a limited number of actual planetary surfaces within the Solar System where in situ scientific investigation by humans is ever likely to occur. The gravity wells of the outer planets may simply be too deep for return missions from the gas giants to be practical. Nearer to Earth, there are only a limited number of sites where special target resources such as helium 3 (for possible energy generation) can be found in the kind of density that might make an extraction industry viable. And there are only a limited number of sites where strategically valuable resources (such as water ice and the most favourable conditions for solar arrays) can be found in combination. While, overall, the opportunities are immense, the resources which will be available at any given time are exhaustible. In some modified form, familiar terrestrial concerns about ensuring that the use of resources remains sustainable, will continue to apply.

The idea of a 'colonisation' of space is deeply embedded in popular culture, yet it may not be the only or best way to make sense of exploration (and possibly settlement). This is not simply a matter of the problematic historic associations of the concept. Rather, it may also suggest that human activity in space is about one single project, rather that a cluster of different projects and activities, involving multiple interests and stakeholders. When the latter are faced with a limited resource, or with limited opportunities, some way of building a reasonable consensus about how to proceed will be required. Such a consensus will have to factor-in (i) the importance of not rapidly exhausting what lies nearest; (ii) considerations about terrestrial impact (e.g. how an emerging space economy with comparatively rapid growth rates will impact upon economic development on the Earth); and (iii) the importance of sustaining continuous access to suitable opportunities for scientific research. For example, on some occasions the window of opportunity to conduct an assessment of the potential astrobiological importance of a site may be restricted, in which case advance preparation (e.g. through work at analogue sites) will be of key importance.

While sites of special scientific or cultural significance (e.g. first-landing sites) may need comprehensive protection from commercial use, many locations will be candidates for both scientific examination and commercial activity. Protocols for site evaluation will need to be developed, and ways of assessing impact devised. As part of a broader overview, there is also a need for new methods to help classify site types and to quantify overall human activity (of various sorts) in order to avoid the problems associated with exponential growth.

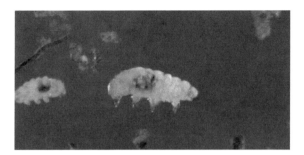

Fig. 5.1 Tardigrades, also known as water bears, are examples of extremophiles, that is, organisms that thrive in ecosystems where at least one physical parameter is close to the known limits for life. *Copyright* Ingemar Jönsson, Kristianstad University, Sweden. Retrieved from the author

5.3 Earth Sites of Importance for Astrobiology

On Earth, three types of environment are especially important sites for astrobiology: habitats for extremophiles (Fig. 5.1); areas that can be used as analogues for environments (and paleoenvironments) on other worlds; and areas that are dark, have a large number of cloud-free nights and a stable atmosphere and are therefore suitable for space observation. These areas match up with three of the deep questions about life introduced in previous sections: How does life begin and evolve? Does life exist elsewhere in the universe? What is the future of life on Earth and beyond? Analogues are of particular importance. Space exploration and research confirms that the Earth is a system, but also a model. New discoveries on other moons and planets (e.g. Mars) are largely supported and guided by knowledge obtained from the study of particular analogue sites on Earth. These planetary analogues have distinctive geological and environmental features which approximate, in specific ways, those we may encounter on other celestial bodies in their present condition, or which may have prevailed during their earlier geological histories. Here, we may think of the Pilbara region in Australia, an ancient seabed with evidence of some of the earliest microbial life. We may think also of the Barberton area in South Africa, which contains some of the Earth's oldest mountains, and evidence of early Earth conditions.

Such analogues are especially important not only for samples but for preparing missions to other worlds through the gathering of general knowledge, the testing of equipment and (in the case of manned missions) the training of the crews. For example the ESA-Pangaea course (Planetary ANalogue Geological and Astrobiological Exercise for Astronauts) in Lanzarote is designed to provide European astronauts with the practical knowledge required to work as effective partners for planetary scientists and engineers.

Astrobiology can reshape our understanding of which resources are scarce, or more scarce than we might imagine. With regard to the third type of site that is of importance to astrobiology, it is significant to note that darkness is a resource that is becoming rare and needs special protection. Some areas that are important for one or more of these three reasons are already protected as world heritage sites, national parks, or UNESCO Global Geoparks. However, in order to be able to continue European astrobiological research, much more has to be done in this respect. It is important to note that new planetary discoveries enrich, through a feedback process, our appreciation of the relevant features of such terrestrial sites.

It is important to deal with the protection and preservation of all three types of sites on a societal and ethical level in order to handle potential conflicts between astrobiological research and other uses. As some of the most important sites for European research are located outside Europe, the process can require complex methodological protocols and multidisciplinary input from Political Science, Applied Ethics and Law.

5.4   Current Regulations and Planetary Protection

The first concerns about contamination that could jeopardize future scientific exploration were raised in 1958 and the first code of conduct to avoid such problems was used by the US Ranger missions to the Moon in 1961. Since then, various missions beyond Earth have orbit have had to implement planetary protection measures of different degrees—ranging from simple documentation to terminal sterilization of entire flight systems. The basic goals for planetary protection regulations are to:

- Ensure that scientific investigations of possible extraterrestrial life forms, precursors, and remnants are not compromised (forward planetary protection)
- Protect the Earth from potential harmful extraterrestrial biological contamination (backward planetary protection).

The forward planetary protection regulations provide the right boundary conditions to allow astrobiological science missions to other moons and planets. The legal basis for planetary protection was subsequently established in Article IX of the 1967 United Nations Treaty on Principles Governing the Activities of States in the Exploration and Use of Outer Space, including the Moon and other Celestial Bodies (Outer Space Treaty). The Committee on Space Research (COSPAR) of the International Council of Scientific Unions (ICSU) maintains the planetary protection regulations as a reference for spacefaring nations and to guide compliance with Article IX of the Outer Space Treaty.

The methodological and legal principles which have shaped planetary protection regulations and measures currently seem sufficiently reasonable and consistent with the science of astrobiology. There are, however, two aspects that require more attention and action by some stakeholders:

- The increased interest of non-governmental entities in space exploration and utilization has shed some light on a regulatory aspect not properly covered by some States that are signatories of the Outer Space Treaty. Article VI of the Treaty clearly states that the States Parties to the Treaty bear international responsibility

for their national space activities and are also required to authorize and continuously supervise the activities carried out by non-governmental actors.

- Understanding the origin, evolution and distribution of life in the universe is a
clearly stated objective in the recommendations of various scientific organizations
and is reflected in the different space agencies programs. In addition to the basic
scientific interest, the presence of extraterrestrial life could also present a threat
to human explorers, equipment they use, in situ resource utilisation, and the terrestrial biosphere after their return. In this context planetary protection has to be
understood as an enabling capability to avoid compromising scientific investigations and to support the preparation of subsequent human exploration and use of
outer space.

To address the first issue a number of State Parties to the Treaty—some of them are
new or prospective spacefaring nations, but some of them pioneers in space—still
need to establish or extend their national regulatory regime to be in line with the full
scope of their responsibilities identified in the Treaty.

To address the second issue it is necessary to ensure that the relevant investments
to further develop planetary protection capability are in place, and are harmonized
with the overall exploration roadmaps at national, European and international level.

Acting on both of these issues is considered essential and time critical in order
to ensure that planetary protection measures continue to protect our investment in
space science, exploration and utilisation of space, for all humankind.

5.5   Environmental Ethics in Space

One of the dominant ethical issues in contemporary astrobiology is the avoidance of
contamination. Such avoidance is partly dealt with by the planetary protection guidelines (described above in 5.4). Planetary protection has been an important aspect of
astrobiology since its emergence in the 1950s, and a key part of the legislative framework for exploration set out in the Outer Space Treaty (OST). As noted previously,
the Treaty does not cover all aspects of contamination since it centers upon worries
about forward contamination (preventing damage to sites in order to protect opportunities for science) and back contamination (ensuring that dangerous materials are not
returned to Earth). This is the legally underpinned framework within which COSPAR
currently makes its recommendations. In recent years there has been a good deal of
discussion, inside and outside of COSPAR, about how the OST can be read in the
'best light' and about whether or not planetary protection should now be thought of
in broader terms than those of Article IX which deals with contamination avoidance.
Also, there has been a good deal of discussion about whether the ethical principles
underlying concern about back contamination should also carry implications for forward contamination. (In which case, they may have to be understood in some broader
way).

It is not obvious that the original focus on protection for scientific examination, and for protecting the Earth from extraterrestrial biological contamination, is sufficiently broad to match up with a more contemporary understanding of environmental ethics. Within the latter, other considerations are regularly brought into play:

- Recognition of the possible value and moral status of extraterrestrial life and of extraterrestrial environments beyond their value as study objects.
- Recognition of the special cultural standing of particular places.
- Sustainability over multiple generations, e.g. protection of places for future generations of humans to experience and to use.
- Wilderness zoning.
- Memorialization of historic sites (e.g. Apollo landing sites).
- Balancing out economic development, scientific interest and ethically driven protection.
- The identification of potentially scarce resources (such as water ice) and how access to, and the use of, such resources is best approached.
- Engaging with ethical and religious discussions about the importance of 'place' and the 'integrity' of locations, as well as discussions about notions of 'inherent value' (beyond human use) that places might have.

To begin to meet these challenges, there is a need for more research in the astrobiological humanities and especially into the crossovers between discussions of space exploration within disciplines such as ethics, law, policy and anthropology. There is also a need for a framework to promote greater engagement between those working within such disciplines and stakeholders within the emerging space economy.

Chapter 6
Education, Training and Scholarship

W. Geppert, D. Dunér, E. Hemminger, Z. Kaňuchová and M. Waltemathe

6.1 Introduction

Astrobiology deals with research questions which are at the core of our human existence. This means that such questions are well suited to attract attention by scholars and students of all age groups, as well as the general public. As it is an interdisciplinary research field, astrobiology can be included as a subject in different STEM disciplines, and in different branches of the humanities and social sciences, at almost all levels of education. In studies involving the use of the Internet, primary school students have already shown an increasing interest in astrobiology-related topics. These include topics such as the identification of habitable exoplanets, icy satellites and minor bodies inside the Solar System. Other topics which have been particularly of interest include planetary missions to look for organic molecules and liquid water on Mars. At secondary school level, astrobiology could provide an integrative framework for forging connections between different subjects and encouraging cooperation between teachers. It could also form the basis for a more project and problem oriented education in science subjects.

W. Geppert (✉)
Department of Physics, Stockholm University, Stockholm, Sweden
e-mail: wgeppert@fysik.su.se

D. Dunér
Lunds universitet, Lund, Sweden

E. Hemminger
Evangelische Hochschule Rheinland-Westfalen-Lippe, Bochum, Germany

Z. Kaňuchová
Astronomický ústav SAV (Astronomical Institute SAS), Tatranská Lomnica, Slovakia

M. Waltemathe
Ruhr-Universität Bochum, Bochum, Germany

© The Author(s), under exclusive license to Springer International Publishing AG, part of Springer Nature 2018

K. A. Capova et al. (eds.), *Astrobiology and Society in Europe Today*, SpringerBriefs in Astronomy, https://doi.org/10.1007/978-3-319-96265-8_6

Fig. 6.1 The Summer school "Biosignatures and the Search for Life on Mars", Iceland, July 2016. The summer school was organised in co-operation with scientists involved in the Nordic Network of Astrobiology, the European Union COST Action "Origins and Evolution of Life on Earth and in the Universe" and the Erasmus + Strategic Partnership "European Astrobiology Campus". *Image Credit* Karen Meech

6.2   Astrobiology Training and Education

The effectiveness of this interdisciplinary approach has been shown by several examples of curricula development in the US. Similar initiatives have also emerged in Europe. The astrobiology research community can contribute to education in Europe by providing training for teachers, visits to schools, organising trips to top level laboratories, and by contributing to textbooks and online information resources. Because of the continuously changing content of scientific knowledge, teaching in astrobiology can also support the development of a flexible and open minded, problem oriented approach towards science. Furthermore (as noted in Sect. 4.2), the high profile of astrobiology-related questions makes astrobiology teaching a useful tool for attracting young people to careers in science and technology (Fig. 6.1).

Training in astrobiology at the university level is of particular importance because astrobiology provides fundamental and intriguing research questions for students and early stage researchers, questions that cannot be tackled by any one discipline alone. Multidisciplinary and interdisciplinary training is therefore important to educate researchers capable of working on these issues. The expertise to provide this training is not usually found in single institutions or even in one country. The Nordic Network of Astrobiology and the COST Action "Life-ORIGINS" have already demonstrated the effectiveness of interdisciplinary astrobiology training for future researchers in Europe through the holding of a number of international summer schools for Ph.D. students and early stage researchers. The training has included practical field work at relevant sites. Nevertheless, a coherent and comprehensive European training structure in astrobiology is required. The above mentioned organisations have therefore created the "European Astrobiology Campus".

A continuation of the activities on the European level is necessary to provide state of the art training in astrobiology at all levels. The main tasks should be thought of as:

- Providing interdisciplinary training for students and multidisciplinary courses, held at astrobiologically interesting places, involving practical research work.
- Encouraging participants in summer schools to plan and perform follow-up research work, and training them in the transferable skills that they will require in order to become future research leaders.
- Creating a forum for lecturers to exchange ideas, concepts, and lecture materials as well as holding meetings on astrobiology training (i.e. training the trainers).
- Providing materials for school teachers in astrobiology.
- Coordinating outreach activities in astrobiology at a European level in collaboration with museums and other stakeholders.
- Creating reference works in astrobiology for the scientific community, for students and for the general public (comparable to the Encyclopedia of Astrobiology).
- Holding summer camps for undergraduate students in astrobiology.
- The organisation of meetings for students and early career scientists.

6.3   Astrobiology and Society in European Scholarship

While a considerable body of scholarship on space policy has already been produced, academic research on the societal aspects of astrobiology (with the exception of Space Law) is still in its infancy. As the relevant societal issues are of general interest, they often receive passing mention in scientific publications. However, sustained analysis is comparatively rare in spite of a longstanding recognition of their significance. The very first serious treatment of societal issues in Europe seems to have been in 1974, when the Royal Society of London sponsored a meeting on "the recognition of alien life". Since then, the field has gradually begun to attract serious scholarly and worldwide attention among academic researchers in the social sciences and humanities. European researchers are clearly in the forefront of these efforts, with European research groups tending to take a broader interpretation of "interdisciplinary research" than their US counterparts. A European Astrobiology Institute with a mission of disciplinary inclusiveness would thus take advantage of an important opportunity.

European researchers have been especially active in the last decade. In 2009, the Vatican hosted a week-long conference on extraterrestrial life which brought together scientists and theologians to address social, ethical, and religious issues in astrobiology. The following year, an explicitly multidisciplinary research project began at Lund University on "Astrobiology—Past, Present, and Future". This project involved several humanities researchers and resulted, among other things, in the first international conference on the History and Philosophy of Astrobiology, as well as a special issue of the journal *Astrobiology*, and an anthology published by Cambridge Scholars on the same theme.

Then, in 2011, The Royal Society hosted a conference on "The detection of extraterrestrial life and the consequences for science and society", which resulted in a conference volume. In 2012 The Center for Space and Habitability was founded at Bern University with the explicit mission of fostering dialogue and interactions between the various disciplines interested in broader astrobiological issues. In 2014, the Vatican Observatory cosponsored a conference with the University of Arizona's Steward Observatory on the search for life. In 2015, the Second International conference on History and Philosophy of Astrobiology was organized by the Nordic Network of Astrobiology, and in 2016, COST sponsored the "From Star and Planet Formation to Early Life" conference, which explored various social and ethical themes. There is a solid foundation of collaborative work to build upon.

One of the major challenges for research concerning astrobiology and society is interdisciplinarity, finding opportunities for interdisciplinary/multidisciplinary collaboration both between science and the humanities/social sciences and between disciplines within the humanities/social sciences. A further challenge is integrating relevant research within the humanities/social sciences that can contribute towards research concerning the societal implications of astrobiology.

A future European Astrobiology Institute should take an important role in interdisciplinary and multidisciplinary collaborations for developing an understanding of the societal implications of astrobiology research. It will give a platform for mutually beneficial collaborations across disciplines, and take advantage of contemporary research within the humanities and social sciences which has special relevance for the understanding of astrobiology and society in Europe today.

6.4   Science and Technology Awareness among Students

Research data on the attitudes toward science and technology held by students enrolled in teacher-training programs in the humanities and social sciences shows a dichotomy between attitude and knowledge. The students have only basic knowledge about key concepts of science and technology. At the same time, these same students feel it necessary to ethically and morally evaluate science and technology and their place in our modern society. Also, in spite of a widespread interest in science and technology among university students from the humanities and social sciences, they will often lack key competencies that would enable them to make useful connections between their own disciplines and science and technology research. The reasons for this are complex but include discipline specialization and differing research cultures, as well as the depiction of science and technology in humanities and social science courses.

As a result, social science students may find it hard to justify attempts to address social, philosophical and ethical issues in science and technology. Astrobiology as a discipline would greatly benefit from bridging the gap between the two. Astrobiology uses existing knowledge about life in order to extrapolate into unknown realms. It teaches us how to approach the unknown in an exemplary fashion while at the same time including broader, and sometimes traditional, philosophical, religious and social questions about knowledge and how it is produced. The social sciences and humanities can bring a methodological sophistication to these issues and to the analysis of data, structures and processes, as well as a deepened understanding of the production of knowledge, its social impact, and its place within a larger pattern of historical social development. To do that, social scientists and academics from the humanities need to rely on a thorough understanding of science and technology.

The main challenges in this area are those of promoting greater familiarity with current science and technology on the part of students and researchers from the humanities and social sciences, but also to educate scientists about the methodology of the humanities and social sciences. It is also necessary to develop analyses of underlying assumptions about scientific practices and scientific reasoning within astrobiology. Such analyses will have connections to non-STEM fields and will motivate students and researchers to acquire the competences required to participate, in an informed way, in ongoing discussions about astrobiological knowledge.

There is a need for a closer identification of the role of the social sciences and humanities within an interdisciplinary practice of astrobiology as well as a need for communicating non-STEM methodology and practices to scientists in order to foster a true interdisciplinary dialogue on a level that is acceptable for academic research. Astrobiology lends itself to this task, as it is already a discipline working across the boundaries of traditional academic fields. Care must be taken, however, to ensure that work across boundaries of disciplines is equal in quality on both sides.

Chapter 7
Technological Innovation and Commerce

E. Chatzitheodoridis, K. A. Capova and E. Persson

7.1 Introduction

Technological innovation is partly a result of high-end research. Such research is the outcome of merged efforts between scientists coming from theoretical, experimental, and engineering disciplines when they aim to answer significant open questions. Sometimes, the direct goal is the provision of new services or products, very often with a commercial value. This bringing together of disciplines requires the development of new tools, practices and protocols for scientific research, in order to promote technological development and increase knowledge. The merging of efforts brings together industry, universities, governmental institutions and private organisations. A recent example is provided by the field of nanotechnology, where technological research at universities and institutions has resulted in entrepreneurial ventures and partnerships with existing companies and has produced a wealth of commercial products. We expect that the interdisciplinary and cross-sectoral aspect of astrobiology can also trigger new types of technological innovation. New knowledge in astrobiology will help to generate new problems to be solved, creating a demand for innovative techniques and tools to carry out space research and planetary exploration.

E. Chatzitheodoridis (✉)
School of Mining and Metallurgical Engineering, National Technical University of Athens, 9 Heroon Polutechneiou Street, 15780 Zograton, Athens, Greece
e-mail: eliasch@metal.ntua.gr

K. A. Capova
Durham University, Durham, UK
e-mail: k.a.capova@durham.ac.uk

E. Persson
Lunds Universitet, Lund, Sweden

© The Author(s), under exclusive license to Springer International Publishing AG, part of Springer Nature 2018
K. A. Capova et al. (eds.), *Astrobiology and Society in Europe Today*, SpringerBriefs in Astronomy, https://doi.org/10.1007/978-3-319-96265-8_7

37

7.2 Technological Innovations Driven by Astrobiology

The field of astrobiology provides an excellent example of a mix of a large number of disciplines and sectors. Disciplines such as biology, chemistry, astronomy, geology, planetary science, ecology, physics, material sciences, computer sciences, and medical sciences, join together with ethics and the philosophical, humanistic, and historical sciences to understand the origins of life and how best to search for it. At the same time, engineering fields such as robotics, computer and data engineering, microelectronics, telecommunications, photonics, analytical instruments and laboratory automation, propulsion, and spaceflight engineering, assist in the above endeavour. Multi-analytical techniques gain attention and innovative instruments are produced and tested in terrestrial-analogue environments. Some of these instruments are then employed in astrobiology exploration missions, i.e., missions such as the 2020 mission to Mars, while the same instruments or some others find their way to the market.

Different sectors and stakeholder communities (e.g. research institutions, space agencies, countries, and industries) must cooperate to make space and planetary research missions possible. They jointly expand our reach to greater distances into the cosmos. They advance research methodologies and broaden our knowledge. From this expansion, new ethical guidelines, legislation and regulations emerge, such as those concerning planetary protection. The development of astrobiology, as we know it today, is a dynamic process which is expected to flourish as space missions which involve humans increase in number, and discussions about ways to establish a stable and continuous presence on other planetary bodies become mature dialogues. There is a huge potential for significant economic growth for all communities which have a proactive role in this endeavour.

Focusing on the core of astrobiology, the novel research approach to study the origins, evolution and distribution of life requires new, state-of-the-art, highly miniaturised and hyper-sensitive instrumentation, as well as new analytical protocols. It also requires special ways of processing the new data, and suitable ways of archiving and consolidating this data into knowledge structures that both humans and non-human systems can work with. There is, finally, a demand for more experimental work in the lab, and for field work in extreme environments and within terrestrial analogues. These are some of the challenges that astrobiology has brought forward and already new techniques for high-precision detection and microanalysis, identification of contamination, data reduction, delicate sample handling and storing procedures are being designed.

7.3 Potential Conflicts of Interest

The potential conflicts of interest between astrobiology and the commercial use of space, as well as the need for legal protection of habitable extraterrestrial environments, have become timely topics. In December 2015, President Obama signed the US Commercial Space Launch Competitiveness Act into US law, encouraging commercial exploration and resource utilisation activities such as the mining of asteroids.

In Europe, Luxembourg has taken the lead in this area through its initiative to put together a legal framework for the commercial use of space resources. Several commercial companies already have ambitious plans to utilise outer space, most notably through asteroid mining and space tourism ventures. A number of companies, such as Deep Space Industries and Planetary Resources, have also been formed with the express purpose of mining asteroids. The first 'space ports' for commercial spaceflight are also beginning to emerge across the world, for example in the US, Sweden, and Abu Dhabi, and there has been a sustained discussion about the possibility of establishing a spaceport in the UK.

Both mining and tourism in space are things of the non-distant future. In their early stages, mining operations will only target asteroids, which are of limited interest for astrobiology, and space tourism will initially be restricted to Low Earth Orbit. There are, however, plans to go further, with opportunities then emerging for cooperation, and the potential for conflicts of interest. One key challenge is to make sure that astrobiology and commercial initiatives can manage to coexist peacefully, and cooperate for mutual benefit.

This is a challenge that needs to be handled through open minded discussion and negotiations that will be better facilitated if there is an established body such as a European Astrobiology Institute with the overview, personnel and trust to represent the consensus among researchers. It will need a large input from ethics, law, political science and other social sciences over issues such as data sharing and the co-ordination of priorities driven by research and by commerce.

Chapter 8
Science and Research

N. Mason, K. A. Capova, P. Laine, A. Losiak, Z. Martins, L. Noack
and K. Smith

8.1 Introduction

Astrobiology is one of the new multidisciplinary research fields of the 21st Century requiring collaborations across the physical sciences—astronomers, chemists, geologists, physicists—as well as people from the biosciences and, as discussed in this chapter, the humanities and social sciences. Astrobiology has a along and distinguished history in Europe though only in last decade has it develop its identity as a robust and independent field, one that will be further developed by new structures, such as the European Astrobiology Institute. This chapter looks at how astrobiology will develop in Europe and provide new career paths for next generation of researchers.

N. Mason (✉)
School of Physical Science, The Open University, Milton Keynes MK7 6AA, UK
e-mail: nigel.mason@open.ac.uk

K. A. Capova
Durham University, Durham, UK
e-mail: k.a.capova@durham.ac.uk

P. Laine
Jyväskylän Yliopisto, Jyväskylä, Finland

A. Losiak
Polska Akademia Nauk (Polish Academy of Sciences), Warsaw, Poland

Z. Martins
Instituto Superior Técnico, Universidade de Lisboa, Lisbon, Portugal

L. Noack
Freie Universität Berlin (Free University Berlin), Berlin, Germany

K. Smith
Clemson University, Clemson, USA

© The Author(s), under exclusive license to Springer International Publishing AG, part of Springer Nature 2018

K. A. Capova et al. (eds.), *Astrobiology and Society in Europe Today*, SpringerBriefs in Astronomy, https://doi.org/10.1007/978-3-319-96265-8_8

8.2   Advancing Astrobiology Research in Europe

Several advances are set to transform our understanding of the universe and of how the Earth fits into its rich diversity: studies of the chemical structure of exoplanet atmospheres; data from next generation large telescopes revealing the chemical complexity of the interstellar medium, as well as shifts in our knowledge of the chemical nature of planet formation. Similarly, studies of extremophiles (which, by mass, may exceed 'normal' life), and more sensitive analytical tools to determine when life first emerged on Earth, will reveal more about how our planet evolved into the rich and diverse habitat that it is today.

On these matters, European research has an important, and distinctive, role to play. The well-earned reputation of the USA and the NASA Astrobiology Institute can sometimes draw attention away from the fact that European science has already played a major role in the emergence of astrobiology research. Astrobiology was not created in the 1950s with the Urey and Miller experiments to simulate the Earth's early atmosphere. Rather, as a field of scientific enquiry, astrobiology began in Europe in the 1920s, with the theories of Oparin and Haldane (and possibly earlier with Darwin's conjecture that life began in a warm little pond of shallow water). Studies of the origins of life, as astrobiology research was broadly entitled before the 1990s, were conducted within mainstream European science departments from 1950 to 1990 with European leaders such as Andre Brack playing a key role. Such research predates the idea of a cohesive discipline of 'Astrobiology'. However, the simultaneous discovery of the Allan Hills Martian Meteorite with its putative fossilised bacteria, and the detection of the first exoplanets in the 1990s, helped to reframe and relaunch the field and so 'Astrobiology', in a more official and recognized sense, was born.

The European scientific research community has, perhaps, been slower than its US counterpart to embrace the term and to recognise the opportunities to reinvigorate and develop astrobiology as a multidisciplinary research field. The first European 'Astrobiology' entitled entity, the European Astrobiology Network Association (EANA) was only created in the new century with the first annual EANA meeting held in Frascati in 2001. These early years did, however, coincide with global austerity and with a reduction in the available research funding that has been particularly difficult for new and multidisciplinary fields.

In 2016 the European community, supported by the European Commission-funded AstRoMap project, published the first authoritative and inclusive European Astrobiology Roadmap providing a European view of astrobiology research. It outlined five key research topics:

Topic 1: Origin and Evolution of Planetary Systems;
Topic 2: Origins of Organic Compounds in Space;
Topic 3: Rock-Water-Carbon Interactions, Organic Synthesis on Earth, and Steps to Life;
Topic 4: Life and Habitability;
Topic 5: Biosignatures as Facilitating Life Detection.

There are clear similarities between these research topics and the research themes of the NAI. There is also a distinctly European dimension based on an evaluation of European skills, facilities and expertise.

European astrobiology needs to establish the structures and secure the financial support that will allow it to develop as a mature and active multidisciplinary research field. The AstRoMap proposed the creation of a European Astrobiology Institute (EAI) "to streamline and optimize the scientific return by using a coordinated infrastructure and funding system" and this should be rapidly established. The EAI should target the European Commission's H2020 programme and its successor Horizon Europe to secure funding that will allow a Europe-wide coordination of astrobiology research, for example through a Research Infrastructure opening European field sites and laboratories to the wider European astrobiology research community. The EAI should also build upon the success of education initiatives, such as the European Astrobiology Campus, to promote multidisciplinary astrobiology teaching and to train a new generation of researchers who are designated astrobiologists. The astrobiology community also needs to engage with Marie Curie Schemes for Innovative Training Networks, and to support the development of Early Career Researchers through Marie Curie Fellowships and European Research Council grants. The EAI should act as a forum for the whole European astrobiology research community and ensure that this new discipline is appropriately recognised and supported by the existing fields from which it has developed, and with which it will collaborate (e.g. astrochemistry, earth sciences, and biochemistry).

8.3   Careers in Astrobiology

Astrobiology is an interdisciplinary research field, with scientists being specialists in diverse areas. A career in astrobiology often starts by taking an undergraduate degree in a specific subject related to astrobiology, e.g. biology, chemistry, physics, geology, astronomy, etc. In contrast to the US and Canada, there is, as yet, no dedicated undergraduate degree offered in astrobiology in Europe. Several European countries, however, provide specific courses or modules at Universities on astrobiology or astrobiology-related topics. Online courses are also becoming increasingly popular. The next step is to do a Ph.D., with a project that is related to astrobiology and/or in a group that performs research in astrobiology. At this point in the career, it may be better to go to one of the universities with research groups focused on astrobiology as this can result in greater experience in the research field, better networking and participation in publications related to the topic.

Careers can be in academia (i.e. University, Museums, Research Institutes), and also in non-academic/science sectors, e.g. industry and space agencies. Most positions in academic institutes are short-term offers. Contracts at space agencies and industry partners, in particular with involvement in space missions, will last as long as a project or mission continues, and therefore there is no long-term perspective, or career progression. Furthermore, research fellowships from ESA to work in astrobiology are included in the Planetary Science topic. National and international funding agencies are more and more open to the topic of astrobiology. This is due in part to the fact that research proposals nowadays often have to be interdisciplinary, and astrobiology fits in this category by definition. To reinforce this, there are increasing numbers of astrobiology-relevant networks which have been funded in recent years in Europe, including European-wide COST networks. ESA has formed different astrobiology-relevant topical teams, leading to several publications on the future of astrobiology in Europe. This has been reinforced by the European Astrobiology Roadmap, AstRoMap, which was supported by the European Commission Seventh Framework Programme.

One of the main challenges of a career in astrobiology is how to attract future astrobiologists. Interested students cannot just study Astrobiology. They need to focus first on a different discipline, and only later transfer to this challenging multidisciplinary field. Furthermore, students do not automatically know which books related to astrobiology to read, both at an undergraduate level and graduate level. Students also do not always know which degrees to apply for in order to build a later career in astrobiology. Online platforms would be a possible solution and a great way to advise students. However, this has been done so far only at national level in a couple of countries, and a Europe-wide platform is still missing.

Another challenge is the fact that astrobiology is sometimes considered to be too broad and lacking a specific focus, which can have implications in terms of funding and/or securing a tenured job. Specifically, when a project is funded for a specific topic, it might have nothing to do with astrobiology. Also, funding is often for only a short period of time (i.e. a couple of years at most after finishing a Ph.D.) unless it is tied to a long-term project. Finally, astrobiology conferences are often not selective enough in their review of abstracts that can harm the way astrobiology is seen by the broader public.

In order to attract new students and researchers to the field, and enhance chances for securing funding, it is necessary to increase awareness of astrobiology as a scientific discipline. This should be done through a broad involvement of all current Astrobiologists in the popularization of astrobiological science. Actions should be focused on two main target groups. The first target group is the general public; basic astrobiological information should be spread through public lectures, articles in magazines/newspapers, and TV and radio appearances. Reaching this group is part of a long-term (i.e. decades-long) strategy of "raising" future students and increasing public willingness to fund astrobiological research. The second target group is current students and researchers of relevant subjects (biology, geology, chemistry, astronomy and physics). Information about astrobiology should be disseminated through seminars and short courses, and by including astrobiological topics within the curriculum or in relevant classes. This is a short term (months to years) strategy to attract new students to the field, and to enhance cooperation between Astrobiologists and representatives of the more established fields such as biology and geology.

Astrobiologists are trained in a specific scientific field, but then need to apply their knowledge in the interdisciplinary field of astrobiology. To smooth the way it would be advantageous to have more astrobiology overview courses or astrobiology relevant courses at national universities. These courses can also be coordinated internationally.

Additional training might be needed for early career scientists to be able to understand the interdisciplinary context of astrobiology, and to avoid misunderstandings by "not speaking the same language" as the other disciplines. For example, training schools including a hands-on session in which people from different disciplines work together on a common research project can help to build bridges between the disciplines. The goal can also be achieved by national and international training and networking symposia, for example the biennial symposium Astrobiology Graduates in Europe (AbGradE) or the Exobiologie Jeunes Chercheurs meeting (EJC) in France, which bring together students and early career scientists from all astrobiology-related fields. However, contributions that are not based on scientific facts or data should ordinarily be rejected, and we need to clearly present this approach to the general scientific community backing the field of astrobiology.

More online tools and initiatives could be developed at an international level, to compare different career paths in astrobiology, summarize web training possibilities, and give an overview of astrobiology-relevant projects at national and international level. This should be a task of a future European Astrobiology Institute. Such an astrobiology forum could also help to foster joint networks and to establish new astrobiology related research programs. Free web-based astrobiology courses (similar to the Coursera program from the University of Edinburgh) could be developed with a focus on bridging between disciplines (for example, including courses such as biology for geologists or geology for physicists).

Training the next generation of astrobiologists in the skills required for writing larger proposals is also important, as it can lead to both establishing a larger, well-linked astrobiology community, and to more joint efforts for project proposals. Coordination of cooperation between different labs and research institutes (for example via a research infrastructure) can contribute to better linking of the different astrobiology disciplines with each other, and to the building of a strong community, which will in turn help to make astrobiology a well-respected field in science. In astrobiology, we need to think bottom-up, as well as institutionally, and start new initiatives for the future. The next generation of astrobiologists needs to be well equipped for that process, to ensure more successful careers in astrobiology.

8.4   The Social Study of Astrobiology as a Science

Astrobiology presents new directions for social study. For example, epistemological cal issues emerge concerning the status of astrobiology as a science, the problematic nature of inference from fragmentary and non-representative evidence, and the limits of scientific enquiry. Perhaps the most important (and engaging) aspect of astrobiology is the study of life's origins and the possibility of life beyond Earth. We need to know where we came from and who might inhabit the universe with us to better understand the cosmos and humankind's place within it.

Topics of enquiry in the social and conceptual areas include the grand metaphysical and scientific issues, such as how best to define "life", "mind", "intelligence", and "culture". Of course, each of these terms has been extensively debated, but almost entirely in a terrestrial, human-centered context. Additionally, the role of contingency and necessity in the origin of these fundamental phenomena require investigation if we are to understand whether or not life, and in particular intelligent life, is likely to be a regular product of natural processes, a unique exception, or something in between. Social sciences and humanities need to recognise the importance of astrobiology in forming these questions as well as in shaping the answers. One challenge is the development of methodology for conducting ongoing and dynamic social research to advance our understanding of the impact of astrobiology on society at a European level. Anthropology of science, that focuses on science as situated in its social contexts and takes into account everyday dimensions of knowledge production, would provide useful assistance in future studies of social aspects of astrobiology.

It is proposed that the social study of astrobiology follow three initial directions. Firstly, a focus upon societal issues such as the likely theological, ethical, and worldview impact of the discovery of microbial or intelligent life; along with an examination of whether the search for extraterrestrial life should be pursued at all, and with what precautions. Secondly, a focus on issues related to the sociology of scientific knowledge, including the diverse attitudes and assumptions of different scientific communities and different cultures towards the problem of life beyond Earth. This includes areas of science and technology studies and social studies of astrobiology. And thirdly, a study of the impact of astrobiology upon society and upon popular beliefs about life beyond Earth, the meaning of life and humanity's place in the universe. The large scale study of public understanding of, and engagement with, science and technology on the European level is a challenging task yet would prove relevant to our understanding of the broader impact of science upon contemporary society and public attitudes to astrobiology alike.

Chapter 9
Leading the Future of Astrobiology in Europe

W. Geppert and M. Gargaud

9.1 Introduction

The advent of the Space Age has put humankind in a position to be able to tackle long established questions about life, its origin and distribution, in a scientific manner. Exciting new discoveries about extrasolar planets, life under extreme conditions, and the existence of hydrothermal systems on the Saturnian satellite Enceladus have fuelled speculations about extraterrestrial life. The detection of complex organic molecules in the interstellar medium, star-forming regions and protoplanetary disks also raises the question of a possible extraterrestrial origin of the basic molecular building blocks of life.

It is clear that we are increasingly well placed to answer a number of major questions concerning not only the possibility of extraterrestrial life and the origin of the building blocks of life on Earth. Other questions are also moving within reach, questions concerning the co-evolution of the biosphere and the geosphere on our planet, and the influence of cosmic events upon it (e.g. changes of solar luminosity, as well as meteorite, comet and asteroid impacts). However, these questions cannot be tackled by one branch of science alone. They require intensive cooperation from scientists across disciplines that might otherwise seem unrelated. And the opportunity to answer such questions is emerging at a time when the European science landscape is undergoing a fundamental transition from a discipline-centered, to a more research-question centered, form of organization. New fields have emerged around burning issues of early evolution of life, exoplanet research and astrobiology. These fields areas are inherently multidisciplinary and carry the potential for new

W. Geppert (✉)
Department of Physics, Stockholm University, Stockholm, Sweden
e-mail: wgeppert@fysik.su.se

M. Gargaud
Laboratoire d'Astrophysique de Bordeaux, Bordeaux, France

© The Author(s), under exclusive license to Springer International Publishing AG, part of Springer Nature 2018
K. A. Capova et al. (eds.), *Astrobiology and Society in Europe Today*, SpringerBriefs in Astronomy, https://doi.org/10.1007/978-3-319-96265-8_9

47

forms of interdisciplinary cooperation and thinking across boundaries of scientific subjects.

9.2 Challenges Facing the EU Astrobiology Community

Multidisciplinarity and interdisciplinarity also entail challenges and even risks. Firstly, scientists have to "learn" the language of other disciplines, which might result in some initial misunderstandings and misconceptions. Secondly, interdisciplinary fields require the training of a new generation of scientists who are able to work across science branches and successfully interact and collaborate with colleagues from other and seemingly remote science branches. This is a task that the European higher education system is still only weakly prepared for. The associated difficulties are sometimes increased because it takes time for new research fields to feed into the often very traditional curricula of universities.

Thirdly, there is an inherent risk that pseudoscience may enter interdisciplinary fields, especially when the subject is vast, few researchers can claim full expertise on even parts of it, and the subject enjoys a great deal of public interest. There is even a danger that such fields may attract some individuals whose talent for publicity greatly exceeds their commitment to reliable norms of science. Without proper accreditation, procedures and regulation, incidents of this sort may not only endanger the public standing of astrobiology, but also the broader reputation of the entire scientific community. It is therefore important to create structures in order to ensure impeccable scrutiny and thus the high quality of astrobiology-related research in Europe. A European Astrobiology Institute functioning as a consortium of prestigious research institutions across Europe would be an ideal structure to help accomplish this task. It would also have the intellectual strength to be able to shield science from the encroachment of pseudoscience.

The strength of any scientific community is, in the long run, determined by its ability to encourage the development of the best scientists and to curtail the influence of persuasive imitations of science. Excellence attracts other excellence and, by establishing contacts with eminent researchers and scholars outside their own field, scientists can extend their own scope of research. This, however, depends upon the existence of efficient fora and structures to foster the required kinds of interaction. The NASA Astrobiology Institute (NAI) is a textbook example of an entity that has succeeded in nurturing such a lively research community. The main conference in the field in the US (AbSciCon) now attracts around one thousand researchers. Internationally, leading research groups at leading institutions based in the United States are competing to obtain the status of an NAI team with quality standards set high enough for several research consortia in Ivy League universities to have been turned down. European astrobiology has been lagging behind these institutional developments, but the various networks mentioned above kindle well-founded hopes about its future.

9.3   The Role of the European Astrobiology Institute

The European Astrobiology Institute will act as a strong voice for the astrobiology community in dialogue with decision makers, funding agencies, the media, other stakeholders, learned societies and the general public. Whereas the EAI will be a consortium of institutions, EANA is (and will remain) a society of individuals. The president of EANA is (as are 4 other EANA Executive Council members) a member of the Interim board of the EAI and discussions on the establishment of the EAI have been held at several EANA workshops.

The EAI will be proactive in the debate on important legal and ethical issues in astrobiology and space research.

9.3.1   Principal Tasks

The principal tasks of the EAI should be as follows:

- To promote and implement multidisciplinary European research projects in all fields of astrobiology.
- To foster international collaboration in astrobiology inside Europe as well as together with other astrobiology communities.
- To act as a forum to discuss new findings in astrobiology through the organisation of conferences, meetings, and online seminars, and to disseminate new research highlights among the scientific community and the general interested public in an apt way after thorough scrutinisation.
- To form the kernel of a network for institutions and institutes, as well as for researchers, to plan cooperative astrobiology projects.
- To perform training, education, outreach and result dissemination in astrobiology in Europe in a comprehensive, synergistic and collaborative way.
- To collect and disseminate astrobiology-related scientific and programmatic information.
- To initiate proposals for grants in order to obtain external financial support (e.g. EU projects).
- To engender debates on important legal and ethical issues in astrobiology and space research.
- To interact with European transnational organisations (e.g. ESA) and European Research agencies on programmatic issues, and to ensure awareness of astrobiology research with decision makers by

  - acting as a strong voice for the European astrobiology community;
  - approaching and informing decision makers in governmental and non-governmental organisations at a national, regional and European level in a coordinated manner in order to promote astrobiology research in Europe as transdisciplinary research activity;

Fig. 9.1 Fresh from Earth.
An image shared by ESA
astronaut Thomas Pesquet on
his social media channels.
Copyright ESA/NASA.
Retrieved from: https://goo.g
l/9J8dVN

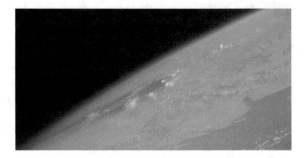

– contributing to the development of a common European Research Area and
  research policy in the field of astrobiology.

• To formulate and continuously update a long-term astrobiology research strategy
  for the European Research Area.
• To collaborate with astrobiology networks and institutes inside Europe and beyond.

While astrobiology, at first glance, does not appear to be a priority subject for industries, there are nonetheless areas of astrobiology that do also attract the interest of
the private sector. These range from the study of enzymes used by extremophiles
and employed in commercial detergents, through to genetic research and work on
space technology. In order to efficiently exploit the possibilities for cooperation and
synergies between research and industry, the European Astrobiology Institute should
create a team for the liaison between research and industry, comprising representatives from both.

9.3.2 Research Areas

The main research areas of the EAI should be carefully defined but inclusive, in order
to avoid leaving important (or newly emerging science areas) outside of the official
reach of research support. A possible list of key areas might include the following
(Fig. 9.1):

• **Formation of planetary systems and detection of habitable planets and moons:**
  How are planetary systems formed? How do the conditions of the formation environments (e.g. at the level of galaxy or protoplanetary disk) influence the formation
  of habitable planets? Which factors define habitability? How can we detect extrasolar habitable planets and satellites?
• **Co-evolution of early Earth's geosphere, atmosphere and biosphere:** How did
  physical, chemical, geological and biological processes co-evolve on Earth? How
  did habitability evolve on early Earth? What conclusions can we draw for other
  planets?

- **Early life and life under extreme conditions:** In which kind of environment did life first emerge (Darwin's little warm pool or some more extreme environment)? What boundary conditions exist for life and what can they tell us about early terrestrial life and the possibility of extraterrestrial life?
- **The pathway to complexity - From simple molecules to first life:** Where and how did the complex organic molecules necessary for life originate (in space, in the atmosphere, or on the surface of the planet) and how were they delivered? How does the environment affect the production and stability of complex organic molecules? How did the formation of biopolymers and self-assembly of first cells proceed?
- **The search for life in early and extreme terrestrial environments and on other planets:** Which strategies should we employ for tracing early terrestrial, as well as extraterrestrial, life in environments? Which combination of individual biosignatures (chemical, geological, spectroscopic, other) and tracers of life present in these environments would be seen conclusive? What novel methods and technologies can be developed to detect life?
- **Historical, philosophical, societal and ethical issues in astrobiology:** How did our ideas about the origin of life develop? What views about extraterrestrial life exist in different cultures? What philosophical, societal, political, juridical and ethical issues are raised by the search for life on other planets and moons?

Research activities in these areas should be coordinated by working groups consisting of active researchers in the field. They can organize special meetings as well as training events, and launch collaborative research projects. The number, focus and themes of the different working groups should be responsive to new developments in astrobiology.

9.3.3 Funding and Mentorship

As in all fields, research in astrobiology is dependent upon access to proper funding. For a thriving research community, it is indispensable to approach funding agencies in a concerted and organized manner to make the maximum use of available grants. The EAI will be well placed to ensure that all possible funding avenues are used by the community and that duplicate approaches are avoided. It will also be able to approach funding agencies in a proactive manner, in order to include grants for interdisciplinary research into funding programmes. Furthermore, the institute will work to introduce astrobiology questions into the scientific programme of space missions at an early stage of planning. The Institute can also work as an information hub to facilitate access to large research infrastructures and field sites. Access to both usually involves extensive red tape which can mean that important resources are still not utilised by the astrobiology community to a satisfactory extent. A centrally maintained and continuously updated information website run by the EAI would be invaluable as a means to encourage more scientists to use the available facilities and to perform field work at astrobiologically relevant sites. This holds especially for early career scientists. The summer schools organized by the Nordic Network of Astrobiology and the European Astrobiology Campus have been very successful in launching the scientific projects of students and young scientists, projects which have originated in field work performed during training events.

The EAI will also be well-placed to develop a mentorship programme for students and for early career scientists who are starting out on their own research projects. A comparable mentorship programme has been successfully initiated by the SAGANet astrobiology grassroots organization in the US. Finally, the Institute can provide an ideal forum for scientists to meet and start new consortia to launch new research projects. Such efforts can range from smaller scale studies through to the proposal of actual space missions.

9.3.4 Suggested Activities of the EAI

Although it brings together many established disciplines, astrobiology itself is a comparatively new research field and so it has to firmly establish itself within the scientific community. As part of this process, the following activities should be envisaged:

- The holding of high-level general conferences in astrobiology as well as smaller workshops on specific subjects in order to bring together scientists from different disciplines.
- The formulation and continuous updating of long-term planning for astrobiology research in the European Research Area.
- The provision of a forum for heads of labs/institutes to plan cooperative astrobiology projects.
- The production of high quality reference works.
- The holding of high-level training events and production of materials for training and education on all levels.
- The organisation of web-streamed seminars by leading scientists in astrobiology, and the provision of web based tools to collect and share existing astrobiology lectures of suitably high quality; see for example what has been done recently by the IAU working group on Education and Training in Astrobiology (led by M. Gargaud, France): http://astrobiovideo.com/en/.
- The facilitation of ongoing dialogue between researchers in the sciences, social sciences and humanities regarding conceptual, societal, ethical and existential questions in relation to astrobiology.

Activities of this sort will also serve to attract world leaders to the field – including those who do not yet define themselves as astrobiologists. It is also noteworthy that, as a new research field, astrobiology has already been able to attract a considerable number of female scientists from a variety of associated disciplines, thereby avoiding familiar problems of gender imbalance (and the detrimental impact upon research that this can produce).

9.3.5 *Multidisciplinarity and Networking*

As the subject area of astrobiology is not yet on the radar screen of directors of studies and other training programmes designers, a European-wide effort is necessary. Given the success of the training programme provided by the European Astrobiology Campus in conjunction with the COST Action "Life-Origins", it is suggested that the European Astrobiology Campus continue its work as the principal training entity of the European Astrobiology Institute and that it coordinates training and education activities through the following four key activities:

- Providing a comprehensive multidisciplinary European astrobiology training programme for students and early career scientists, offering both basic and specialised training events in the field as well as training in generic skills such as proposal writing and the planning of scientific projects.
- Creating a network for astrobiology education and providing training materials for basic, secondary and higher education in Europe.
- Providing a mentoring scheme for early career scientists in the field of astrobiology.
- Ensuring proper funding for these activities.

These measures will ensure proper training of the next generation of astrobiologists.

Chapter 10
Afterword

T. Milligan, K. A. Capova, D. Dunér and E. Persson

Space exploration and the search for a better understanding of life have never been entirely separate from one another. This is not simply a matter of policy, a decision by political administrations to combine the two. Rather, it is a matter of the ways in which both draw upon the same scientific culture and upon overlapping societal influences. Some of the latter are the political influences of particular times and particular places, others are of a far broader nature. Progress in one field has tended to be combined with advances in the other. It is a familiar point that, in the very year that NASA was founded, i.e. 1958, the American molecular biologist Joshua Lederberg won a Nobel Prize for his discovery that bacteria can exchange genetic material (a process now known as 'bacterial conjugation'). This, in turn, was only six years after the classic Miller-Urey experiment to replicate the production of some of the chemical precursors of life. And the Miller-Urey experiment, in turn, overlapped in time with the work of Watson, Crick and Rosalind Franklin in England, on the structure of DNA. Major breakthroughs came in both fields (activity in space and research into life) within the same time-frame, and drew upon at least some of the same background influences and interests. When Lederberg went on to work with NASA on the early programs to look for life on Mars, the progression was, in an everyday sense, natural. Interest in space and interest in life went together. They have always tended to do so.

T. Milligan (✉)
Department of Theology and Religious Studies, King's College London, Room 3.42 Virginia Woolf Building, 22 Kingsway, London WC2B 6LE, UK
e-mail: Anthony.milligan@kcl.ac.uk

K. A. Capova
Durham University, Durham, UK
e-mail: k.a.capova@durham.ac.uk

D. Dunér · E. Persson
Lunds Universitet, Lund, Sweden

© The Author(s), under exclusive license to Springer International Publishing AG, part of Springer Nature 2018
K. A. Capova et al. (eds.), *Astrobiology and Society in Europe Today*, SpringerBriefs in Astronomy, https://doi.org/10.1007/978-3-319-96265-8_10

What motivates both, and justifies a good deal of scientific expenditure at a state and international level, is an overlapping set of human concerns: about where we come from, where we might go and whether or not we can assume that life exists only here. The familiar formal definition of astrobiology, the definition used in the White Paper above ('the scientific study of the origins, evolution, and distribution of life') reflects these concerns and provides a more precise framework for research and for collaborative activities. How quickly this research, and these activities, might proceed is a different matter. In order to be effective over more than the short term, proposals for institutional initiatives which are geared to significantly advance research must be not only 'good in principle', but also timely. Good ideas at inconvenient or impractical moments are not necessarily a solid foundation for research structures. Timeliness matters.

The recent history of our human activities in space can help to illustrate the point. It is littered with initiatives that have produced a good deal of debate about precisely this issue of when actions ought to be carried out. Premature moves do not always end well. Notoriously, Bernard Lovell, the long-term leader of the Jodrell Bank telescope team, described the US Air Force Westford program as 'ethically wrong' for having sent 350 million copper needles into space in October 1961. The intention was that they might act as dipole antennae, in the hope of improving military communications. More needles followed in 1963. In retrospect, it is difficult to disagree with Lovell's assessment. If we had known then what we know now about space debris, it is highly unlikely that this would have been done even once. Moving too quickly, and then repeating the exercise without due caution, resulted in a counterproductive outcome.

Even the Apollo program has been subject to questions along these same lines. The questioning is not so much about the many accomplishments of the program, or even the validity of its core goals, but about its pace and the problem of sustainability. The early US space program was, after all, driven in part by a transitory Cold War, as well as by human imperatives to explore. While it was broadly welcomed, and rightly celebrated for its remarkable achievements, even at the time some public figures such as J. G. Ballard questioned the wisdom of trying to press too far too quickly, for what might ultimately turn out to be a premature space age. Whether or not more would have been done, or could have been done, in the 1970s and 1980s, if the pace of the space program expansion in the earlier 1960s had been more *even* remains a matter of debate. These are issues for the historians, issues that are likely to continue to be debated and will not be settled here. What is done is done. However, the background assumptions, that sustainability matters and timeliness cannot be ignored, are harder to set aside. There is a strong case for accepting that major institutional proposals do require justification, not simply in terms of the validity of their goals but also in terms of both of these key factors: the sustainability and timeliness of the proposals, their prospects for a positive longer term contribution to research outcomes.

The White Paper has made a case for the timeliness of a European Astrobiology Institute. This would be a major move, contributing to cohesion within the European Research Area. In this particular case, the justifications do seem to be strong. First, the proposal comes at a point in time when our human activities in space are significantly expanding and when further and rapid growth in the near future is reasonably anticipated, driven in part by the emerging commercial space sector and its ambitions. Development here is not simply a matter of increasing scale, but also broadening involvement, with both state and private sectors taking joint and separate initiatives, and with Europe as well as the US and Asia gearing up for major steps forward. (These three seem particularly important for the next phases of development.) In the European case, at the time of writing, the European Space Agency in collaboration with the Russian space agency, Roscosmos, is gearing up to send the ExoMars rover to the red planet in 2021 in order to look for atmospheric gasses that might be linked to active geological or biological processes. The rover will be the first to drill down into the surface of the planet, down as far as 2 m. (With all of the hopes for new information and issues of planetary protection that raises.) The move involves an ambition to emulate successes by the US, but also to go further and deeper.

Europe is also a major area for the production of cubesats and is in the final stages of the completion of Galileo, Europe's own global navigation satellite system, comparable to GPS and GLONASS (the US and Russian navigation equivalents). When completed, in or around 2020, it should consist of 24 operational satellites and a further 6 spare satellites positioned in three Medium Earth Orbit planes at 23,222 km above the Earth. Again, the project is ambitious. Similarly, while Europe already has a spaceport in French Guiana, the prospect of future space tourism has led to discussions about the modest development of spaceports within Europe itself. A change that will be hard to avoid once the industry has moved from basic infrastructure and logistics to regular operations.

By comparison, during the Apollo program in the 1960s, although many individual European scientists and technicians were involved, Western Europe was largely an institutional bystander and the European Space Agency had yet to be formed. (Activities were still coordinated by the precursor body, the European Space Research Organisation.) While US space activities still remain significantly larger, Europe is now a major player within an emerging range of commercial and state space activities, with the prospect of playing a full part in the next phase of human activities in space. Seen in the light of this, the proposal for the formation of a European Astrobiology Institute is geared to allow the development of astrobiology in Europe to keep pace with other developments across various space sectors within Europe and elsewhere.

Second, astrobiology now has a strong international research community, and a particularly strong research community within Europe. This is an area of special and sustained European strength. Research does not depend upon a small number of scholars, but upon a sustainable international research community. By comparison with some of the classic disciplines of science, astrobiology is a new research field. However, it is no longer in its infancy. We have come a very long way from Darwin's speculations about life originating in a 'warm little pond', and from the conjectures about life's origins by the European scholars Oparin and Haldane in the 1920s. We also know a good deal about how quickly a new field of scientific research can expand, when it is driven by social considerations as well as scientific curiosity and the search for knowledge. Genetics is an obvious example and one which is, again, linked to the goal of arriving at a better understanding of life as such. The thought that Europe should not play 'catch up' when it can be at the forefront of this research, plays an important justifying role. Given an expanding research field, the questions of 'How integral can work across the European Research Area be?' and 'How integral *should* it be?' are important policy issues. The White Paper has argued that Europe can and should play a major role, that it should be at the center of this research because it has something distinctive to add. It has a distinctive voice, exemplified in the extensive collaboration involved in the production of the White Paper itself. It can speak for a greater culture of co-operation between the sciences and relevant discipline areas within the humanities. More importantly, it can show how such co-operation may proceed.

Finally, in addition to the point that human activities in space are rapidly increasing (and Europe is already part of this expansion), and the point that the European contribution to astrobiology is already significant and distinct, it should be recognized that our knowledge of where to look for life, and where it is unlikely to be found, is far greater now than in even the recent past. As an example, and one which few people are aware of, NASA no longer comprehensively sterilizes landing craft for the surface of Mars. Instead, rovers are constructed in clean rooms with partial sterilization for the simple reason that landing on the surface of Mars, with the exposure to ultraviolet radiation that this involves, is itself an efficient means of sterilization for at least some purposes. (Matters change when, as with ExoMars, there is any prospect of drilling into the surface. Planetary protection then requires further precautions.)

The whole project of exploring the conditions for life, and the possibility of finding traces or life signatures, is vastly more sophisticated than it was during the pioneering days of lander-based exploration. When the Viking landers touched down upon the surface of Mars in 1976, the first extremophile organisms had only recently been discovered here on Earth. Learning where to look and how best to look for *traces* of life was a new challenge. The modelling for the experiments, guided by what was known at the time, was rudimentary, resulting in a false positive and subsequent controversy. By contrast, if there is life elsewhere in the Solar System, we are now far better equipped (both technologically, and in terms of our overall knowledge about life) to find it. It is even tempting to think of ours as an era of 'preparing for discovery' rather than merely one of searching hopefully for life.

Of these three considerations, the first two may turn out to be the more important. The third is, perhaps, more of an ancillary consideration. Important, but it functions largely as a reinforcer of the other two. We do not, ultimately, need any shift to a 'preparing for discovery' setting in order for the two main justifications to stand. They hold up in their own right. The validity and timeliness of the proposal for Europe to further coordinate its leading research role, and give greater cohesion to research activities across the European Research Area, does not depend upon the discovery of life elsewhere. Nor does it depend upon how close we might be to such a discovery. Rather, it turns upon more Earthly concerns. Understanding life on Earth is a scientific and social obligation, an ethical imperative and a practical necessity which reaches beyond any institutional proposals and frameworks. A consensus about the importance of understanding life is at the core of the environmental concern now shared across the world by governments, non-governmental agencies and citizens alike. It is also at the core of the major international agreements whose goal it is to safeguard life, and to protect life on Earth from the worst effects of climate change. But understanding life on Earth, and doing so in a sufficiently deep and detailed way, requires us to consider why life has emerged here yet not in various other places, and why some locations elsewhere might turn out to be better candidates for life (or for historic traces of life) than others. Overall, we know vastly more about these matters than we did in the middle of the last century, when the first space programs emerged. And we will know vastly more again in fifty years' time. This is a period of rapid development in our knowledge and in our grasp of the larger story within which humans are situated.

Consequently, the conducting of the relevant research, the protecting of the relevant locations, and the theorizing of results, has a value that is independent of the discovery of any actual second location where life might once have had a foothold. (Fascinating and important though such a discovery would be.) Astrobiology is, in brief, integral to the deepening of our best understanding of human and terrestrial life. Its defined focus, upon 'origins', 'evolution' and 'distribution', directs our attention towards precisely the range of questions that we need to address if life anywhere (including here on Earth) is to be understood in a way that truly deserves to be called deep and sufficiently detailed. The rapid growth of research into extremophiles is an example. Once regarded as something of a special exception, the succession of discoveries of organisms capable of surviving at extremely hot or extremely cold temperatures, or in conditions of high acidity or alkalinity, is now regarded as integral to our grasp of the resilience and variability of life itself. Hydrothermal vents, the sub-ice waters in Antarctica, and the Marianas Trench (the very deepest place on the surface of the Earth) have been found to harbour life. Automatically discounting the possibility of life securing a hold in difficult and remote locations, is no longer plausible. Coming to grips with the extensive presence of extremophiles on Earth has reshaped scientific narratives. Yet, if life exists in such places here, but not in favourable locations elsewhere, this itself poses questions about its origins and evolution as well as distribution.

Extremophiles are, however, only one example of this entanglement of research into terrestrial life and exploration of the possibilities for life elsewhere. Again, it is the *possibilities* for life elsewhere, rather than any act of discovery, that ultimately seems to matter. The most important advances for our understanding of life are not always the most dramatic. Hence, the challenges of science outreach which the White Paper have drawn attention to. While it is true that a discovery of life traces on another planet, or in a meteor, or even on an asteroid, would be a newsworthy story for the ages, it is simply not a required justification for extensive and detailed astrobiology research. In line with this, speculation about whether or not life will ultimately be found elsewhere has remained outwith the bounds of the White Paper discussion. Not because it is uninteresting. It is obviously an interesting question and a great deal can be said about what we should expect the outcome of such a search to be. Rather, it has remained outwith the limits of the White Paper discussion because of a concern to remain within the bounds of what we actually *know* and can predictably work with.

The thought has not been to make policy suggestions based upon conjecture, no matter how interesting that conjecture might happen to be. Instead, the thought has been to work upon the basis of what is already known, what is understood about how science evolves, and what is reasonably anticipated about future space-related research and activities. What is known, understood, and reasonably anticipated is that the research field of astrobiology will continue to grow; the commercial space sector will expand considerably; Europe and Asia will both be significant players alongside the US; and multiple stakeholders are likely to establish a growing presence in space. For most policy makers, these are non-controversial assumptions.

As a final point, with growing human activity in space comes the inevitable set of discussions about regulations, systems of co-ordination, responsibility, legal and ethical duties and entitlements. In brief, all of the problems of establishing a shared legal framework for international activities by multiple agencies from different nations. Such discussions might issue, at some point, in the revision of existing agreements or in a new international agreement comparable to the Outer Space Treaty of 1967. Or, in the light of the considerable difficulties of securing a new agreement to accommodate the interests of so many legitimate stakeholders, there may be a move to align norms and practical working understandings without any new and overarching space treaty. This too is a matter of conjecture. Reasonable opinions among policy analysts may differ, and especially so in light of the fact that regulation is thought of differently in the US, Europe and across various and distinct parts of Asia. As yet, the outcome of such dialogue remains unclear. That it will take place is not, however, conjectural. It has already begun. And a strong European voice in the field of astrobiology, expressing a broad scientific and scholarly consensus, promises to be an important contributor.

Appendix

A.1 Definitions (Drawing upon the Encyclopedia of Astrobiology)

Astrobiology

Astrobiology is the scientific study of the origins, evolution, and distribution of life, not only on Earth, but also at the scale of the Universe. Concepts such as astrobiology, exobiology, bioastronomy are associated with this scientific activity. They help to define the study of the origins of life on Earth, and elsewhere. They also help to define the search for the existence of life in the Universe. The idea of life itself is represented by 'bio-' or 'biology', and the universal field of study by 'astro-', '-astronomy', or 'exo-'. Beyond the terminological choices, these words suggest a great ambition for contemporary human knowledge, a need to exceed disciplinary divisions.

Extremophiles

Extremophiles are organisms that thrive in ecosystems where at least one physical parameter is close to the known limits for life. While some organisms may temporarily survive harsh conditions by forming resistant stages (spores), or through specific mechanisms (heavy metal resistance), true extremophiles require these harsh conditions. For instance, hyperthermophiles successfully complete their life cycle at optimal temperatures above 80 °C, and commonly do not grow at all below 60–70 °C.

Exoplanets

An exoplanet, or extrasolar planet, is a planet in orbit around another star. The first such object was discovered in 1992 and more than 3700 exoplanets had been identified by April 2018.

© The Author(s), under exclusive license to Springer International Publishing AG, part of Springer Nature 2018

K. A. Capova et al. (eds.), *Astrobiology and Society in Europe Today*, SpringerBriefs in Astronomy, https://doi.org/10.1007/978-3-319-96265-8

Habitability

Habitability for the Solar System refers to suitability for carbon-based life. The presence of liquid water is the first potential indicator of the habitability of any Solar System body. The study of conditions on these bodies is basic to understanding the difference between a habitable planet and one that is not suitable for life.

Planetary protection

Planetary protection is the process of preventing contamination of planetary environments by living protection of organisms from other planets, in accordance with Article IX of the 1967 Outer Space Treaty, and in accordance with policies maintained by COSPAR. Nations sending missions to other planets must ensure that Earth life does not contaminate these planets, and that any samples brought back to Earth do not release harmful organisms into our environment.

A.2 Presentation of the Group

Life-ORIGINS (TD1308) is a Trans Domain European COST Action dedicated to the scientific investigation of the origins and evolution of life on Earth and the habitability of other planets. Working Group 5: History and Philosophy of Science is led by D. Dunér (Sweden; 2014–2018), co-led by Ch. Malaterre (Canada; 2014–2016) and S. Tirard (France; 2016–2018).

The Group's Objectives: Life's nature is strongly dynamic and systemic, forming a collection of interacting parts. It now appears likely that its appearance on Earth is to be explained through the coupling of different types of molecular structures and of evolutionary processes in which these structures are involved. Within this broad scientific context, the overall objectives of this working group cover two main areas of investigation:

- An assessment, from a philosophical perspective, of how the boundaries between chemistry and biology are being transformed as a result of a shift towards increasingly systemic or holistic approaches in the quest for a naturalist explanation of the origin of life. The focus here will be on characterizing the new tools, models and conceptions that are being elaborated to tackle the problem of origins within this wider evolutionary scenario, and on how the latter will provide a completely new view of the chemistry-biology interface.
- An analysis, from a history of science perspective, of the debate concerning the origin of life and the chemistry/biology divide. This will involve study of changes in the definition of the life concept, in the classification and categorization of life, and in models for the evolution of life.

A.3 TD1308 in Images

A.3.1 TD1308 ORIGINS Conference Photograph from Porto, Portugal, 2015

Image Credit Laurence Honnorat, Innovaxium

The 1st TD1308 ORIGINS Conference 'Habitability in the Universe: From the Early Earth to Exoplanets' was held in Porto, Portugal, 22–27 March 2015. The Porto conference was the first Conference and WGPP (Working Group and Project Planning) meeting of the TD1308 COST action ORIGINS. TD1308 European action, gathers 30 countries and 150 scientists working in astrophysics, astrochemistry, planetology, geology, geochemistry, biology, paleontology, space science, engineering, philosophy and history of science, is divided into five working groups, among which 3 of them had their first WG meeting in Porto. Conference Website: http://www.iastro.pt/research/conferences/life-origins2015.

A.3.2. Astrobiology Summer School, 2016

Image Credit Karen Meech

The Summer school "Biosignatures and the Search for Life on Mars" took place in Iceland, 4–16 July 2016. The summer school was organised by Wolf D. Geppert (Stockholm University, Sweden) and David Cullen (Cranfield University, UK) and in co-operation with the University of Akureyri and the other scientists involved in the Nordic Network of Astrobiology, the European Union COST Action "Origins and Evolution of Life on Earth and in the Universe" and the Erasmus + Strategic Partnership "European Astrobiology Campus". The Summer School website: http://www.nordicastrobiology.net/Iceland2016/Lecturers.html.

A.3.3 Presenting the Encyclopedia of Astrobiology

Image Credit Muriel Gargaud

The Encyclopedia of Astrobiology serves as the key to a common understanding. Each new or experienced researcher and graduate student in adjacent fields of astrobiology will appreciate this reference work in the quest to understand the big picture. The carefully selected group of active researchers contributing to this work and the expert field editors intend for their contributions, from an internationally comprehensive perspective, to accelerate the interdisciplinary advance of astrobiology.

The Editor in Chief, Muriel Gargaud (second from right), is an enthusiastic and experienced editor who has proven in various projects that she can manage a large number of editors and authors and deliver an excellent publication. William Irvine (first from right) was President of the Commission on Bioastronomy of the International Astronomical Union. His research activity is concentrated in two areas: the chemistry of dense interstellar clouds; and the physics and chemistry of comets.

A.3.4 TD1308 ORIGINS Conference Photograph from Bertinoro, Italy, 2018

The 4th, and last, TD1308 ORIGINS Conference 'Life on Earth and beyond: emergence, survivability, and impact on the environment' was held in Bertinoro, Italy, 19–24 March 2018. The aim of the conference was to gather experts coming

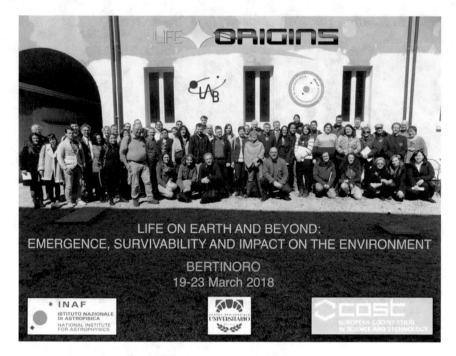

Image Credit INAF—Astrophysical Observatory of Arcetri

from different scientific fields to share ideas, discuss about recent and present results, and identify the key challenges regarding the emergence, survivability, and impact on the environment of life both on Earth and beyond. Conference website: https://www.arcetri.astro.it/ ∼ bertinoro.

A.4 COST Action Memorandum

Memorandum of Understanding for the implementation of a European Concerted Research Action designated as COST Action TD1308: Origins and evolution of life on Earth and in the Universe (ORIGINS). For the implementation of a European Concerted Research Action designated as COST Action TD1308 Origins and evolution of life on Earth and in the Universe (ORIGINS). The Parties to this Memorandum of Understanding, declaring their common intention to participate in the concerted Action referred to above and described in the technical Annex to the Memorandum, have reached the following understanding:

1. The Action will be carried out in accordance with the provisions of document COST 4114/13 "COST Action Management" and document COST 4112/13 "Rules for Participation in and Implementation of COST Activities", or in any

new document amending or replacing them, the contents of which the Parties are fully aware of.

2. The main objective of the Action is to address, using an interdisciplinary approach, three great questions about the origin, evolution and distribution of life: (a) where, when and how did life emerge and evolve on Earth? (b) what are the conditions under which life can exist? (c) does life exist elsewhere in the Universe and, if it does, how can it be detected and identified?

3. The economic dimension of the activities carried out under the Action has been estimated, on the basis of information available during the planning of the Action, at EUR 72 million in 2013 prices.

4. The Memorandum of Understanding will take effect on being accepted by at least five Parties.

5. The Memorandum of Understanding will remain in force for a period of 4 years, calculated from the date of the first meeting of the Management Committee, unless the duration of the Action is modified according to the provisions of section 2. Changes to a COST Action in the document COST 4114/13

Suggested Literature

1. Balla, V. K., Roberson, L. B., O'Connor, G. W., Trigwell, S., Bose, S., & Bandyopadhyay, A. (2013). First demonstration on direct laser fabrication of lunar regolith parts. *Rapid Prototyping Journal, 18,* 451–457.
2. Barnett, M. J., Pawlett, M., Wadham, J. L., Jackson, M., & Cullen, D. C. (2016). Demonstration of a multi-technique approach to assess glacial microbial populations in the field. *Journal of Glaciology, 62,* 348–358.
3. Bedau, M. A., & Cleland, C. E. (2010). *The Nature of Life: Classical and Contemporary Perspectives from Philosophy and Science.* Cambridge: Cambridge University Press.
4. Bertka, C. M. (2009). *Exploring the origin, extent, and future of life: Philosophical, ethical, and theological perspectives. Cambridge astrobiology 4.* Cambridge, UK, New York: Cambridge University Press.
5. Billings, L. (2007). Overview: Ideology, advocacy, and spaceflight. Evolution of a cultural narrative. In J. S. Dick, R. D. Launius (Eds.), *Societal impact of spaceflight* (pp. 483–499). Washington, DC: National Aeronautics and Space Administration Office of External Relations (The NASA history series, 4801).
6. Billings, L. (2012). Astrobiology in culture: The search for extraterrestrial life as 'Science'. *Astrobiology, 12*(10), 966–975.
7. Capova, K. A. (2016). The new space age in the making: Emergence of exo-mining, exo-burials and exo-marketing. *International Journal of Astrobiology, 15*(4), 307–310.
8. Capova, K. A. (2013). The detection of extraterrestrial life: Are we ready? In Vakoch, D. (Ed.), *Astrobiology, history, and society.* Springer.
9. Cockell, C. S. (2001). 'Astrobiology' and the ethics of new science. *Interdisciplinary Science Reviews, 2*(2), 90–96.
10. Cockell, C., & Horneck, G. (2004). A planetary park system for Mars. *Space Policy, 20*(4), 291–295.
11. Consolmagno, G. (2014). *Would you baptize an extraterrestrial? ... and other strange questions from the inbox at the Vatican Observatory* (1st ed.). New York: Image.
12. COST Action TD1308 (2013). Memorandum of understanding for the implementation of a European Concerted Research Action designated as COST Action TD1308: Origins and evolution of life on Earth and in the Universe (ORIGINS). Source: https://e-services.cost.eu/files/domain_files/TDP/Action_TD1308/mou/TD1308-e.pdf.
13. Cottin, H., Kotler, J. M., Bartik, K., Cleaves, H. J., Cockell, C. S., de Vera, J.-P. P., et al. (2017). Astrobiology and the possibility of life on earth and elsewhere.... *Space Science Reviews, 209,* 1–42.
14. Cottin, H., Kotler, J. M., Billi, D., Cockell, C., Demets, R., Ehrenfreund, P., et al. (2017). Space as a tool for astrobiology: Review and recommendations for experimentations in earth orbit and beyond. *Space Science Reviews, 209,* 83–181.
15. Chon-Torres, O. A. (2018). Astrobioethics. *International Journal of Astrobiology, 17*(1), 51–56.

© The Author(s), under exclusive license to Springer International Publishing AG, part of Springer Nature 2018
K. A. Capova et al. (eds.), *Astrobiology and Society in Europe Today*, SpringerBriefs in Astronomy, https://doi.org/10.1007/978-3-319-96265-8

16. Cousins, C. R., & Cockell, C. S. (2015). An ESA roadmap for geobiology in space exploration. *Acta Astronautica, 118,* 286–295.
17. Crowe, M. J. (1999). *The extraterrestrial life debate, 1750–1900.* Mineola, N.Y: Dover Publications.
18. Crowe, M. J. (2008). *The extraterrestrial life debate, Antiquity to 1915: a source book.* Notre Dame IN: University of Notre Dame.
19. Debus, A. (2004). COSPAR needs for planetary protection recommendations for sample preservation dedicated to exobiology. *Advances in Space Research, 34*(11), 2320–2324.
20. Dick, S. J. (1996). *The biological universe: The twentieth-century extraterrestrial life debate and the limits of science.* Cambridge University Press.
21. Dick, S. J. (2012). Critical issues in the history, philosophy, and sociology of astrobiology. *Astrobiology, 12*(10), 906–927.
22. Dick, S. J. (Ed.). (2015). *The impact of discovering life beyond earth.* Cambridge: Cambridge University Press.
23. Dick, S. J. (2018). *Astrobiology, discovery, and societal impact.* Cambridge University Press.
24. Dominik, M., Zarnecki, J. C. (Eds.). (2011). *The detection of extra-terrestrial life and the consequences for science and society* (Vol. 369, No. 1936, pp. 497–699). London: The Royal Society.
25. Dunér, D., Malaterre, Ch., & Geppert, W., (Eds.). (2016). The history and philosophy of the origin of life. *Special issue of International Journal of Astrobiology.*
26. Dunér, D., Persson, E., & Holmberg, G. (Ed.). (2012). The history and philosophy of astrobiology. *Special issue of Astrobiology, 12*(10).
27. Dunér, D., Parthermore, J., Persson, E., Holmberg, G., (Eds.). (2013). *The history and philosophy of astrobiology: Perspectives on extraterrestrial life and the human mind.* Cambridge Scholars.
28. European Cooperation for Space Standardization, ECSS-U-ST-20C Space Sustainability— Planetary Protection.
29. Garcia-Descalzoa, L., García-Lópeza, E., Morenoa, A.-M., Alcazarb, A., Baqueroa, F., & Cid, C. (2012). Mass spectrometry for direct identification of biosignatures and microorganisms in Earth analogs of Mars. *Planetary and Space Science, 72,* 138–145.
30. Gargaud, M., William, M. I., Amils, R., Cleaves, H. J., Pinti, D. L., Quintanilla, J. C., et al. (Eds.). (2015). *Encyclopedia of astrobiology.* Heidelberg: Springer.
31. Harrison, P. (2016). *The myth of a perennial battle between science and religion.* Available online at https://www.theologie-naturwissenschaften.de/startseite/leitartikelarchiv/conflict-myth.html, checked on February 16, 2017.
32. Horneck, G., et al. (2016). AstRoMap European astrobiology roadmap. *Astrobiology, 16*(3), 201–243.
33. Kereszturi, A., Bradak, B., Chatzitheodoridis, E., & Ujvari, G. (2016). Indicators and methods to understand past environments from ExoMars Rover drills. *Origins of Life and Evolution of Biospheres.*
34. Kissel, J., Altwegg, K., Clark, B. C., et al. (2007). Cosima—high resolution time-of-flight secondary ion mass spectrometer for the analysis of cometary dust particles onboard Rosetta. *Space Science Reviews, 128,* 823.
35. Kminek, G. (2015). Planetary protection—An enabling capability and contributor to sustainable space exploration. *Space Research Today, 194,* 5–6.
36. Kminek, G., Conley, C., Hipkin, V., & Yano, H. (2017). COSPAR's planetary protection policy. *Space Research Today, 200,* 12–25.
37. Magnani, P., Re, E., Senese, S., Rizzi, F., Gily, A., & Baglioni P. (2010). The drill and sampling system for the ExoMars Rover. In *i-SAIRAS Conference* (044-2769).
38. Martins, Z., Cottin, H., Kotler, J. M., Carrasco, N., Cockell, C. S., de la Torre Noetzel, R., et al. (2017). Earth as a tool for astrobiology—A European perspective. *Space Science Reviews, 209* (1–4), 43–81.

39. McKay, C. P. (1987). Terraforming: Making an Earth of Mars. *The Planetary Report, 7,* 26–27.
40. Musso, G., Lentini, G., Enrietti, L., Volpe, C., Ambrosio, E-P., Lorusso, M., et al. (2016). Portable on orbit printer 3D: 1st European additive manufacturing machine on international space station. *Advances in Physical Ergonomics and Human Factors* 643–655.
41. NASA Technical Memorandum: Workshop on the Societal Implications of Astrobiology, Final Report, Ames Research Center, November 16–17, 1999.
42. Noack, L., Verseux, C., Serrano, P., Musilova, M., Nauny, P., Samuels, T., et al. (2015). Astrobiology from early-career scientists' perspective. *International Journal of Astrobiology, 14*(4), 533–535.
43. Noffke, N. (2005). Introduction: Geobiology—A holistic scientific discipline. In N. Noffke (Ed.), *Geobiology: Objectives, concepts, perspectives.* Elsevier.
44. Parrish, C. F. (2003). Space colonization. *Aerospace America, 41*(12), 94.
45. Persson, E., Capova, K. A., & Li, Y. (2018). Attitudes towards the scientific search for extra-terrestrial life among Swedish high school and university students. *International Journal of Astrobiology* 1–9.
46. Persson, E. (2013). Philosophical aspects of astrobiology. In D. Dunér, J. Pathermore, E. Persson, G. Holmberg (Eds.), *The history and philosophy of astrobiology* (pp. 29–48). Cambridge Scholars.
47. Persson, E. (2014). What does it take to establish that a world is uninhabited prior to exploitation? A question of ethics as well as science. *Challenges* (5), 224–238.
48. Persson, E. (2017). Ethics and the potential conflicts between astrobiology, planetary protection and commercial use of space. *Challenges, 8*(1), 12.
49. Peters, T. (2011). The implications of the discovery of extra-terrestrial life for religion. *Philosophical transactions A, Mathematical, physical, and engineering sciences, 369*(1936), 644–655.
50. Peters, T., Froehling, F. (2008). *The Peters ETI religious crisis survey.* Source: www.counterbalance.net/etsurv/index-frame.html.
51. Peters, T. (2009). Astrotheology and the ETI myth. *Theology and Science, 7*(1), 3–29.
52. Race, M., et al. (2012). Astrobiology and society: Building an interdisciplinary research community. *Astrobiology, 12*(10), 958–965.
53. Rettberg, P., et al. (2016). Planetary protection and Mars special regions—A suggestion for updating the definition. *Astrobiology, 16,* 119–125.
54. Rochus, P., Plesseria, J. Y., Van Elsen, M., Kruth, J.-P., Carrus, R., & Dormal, T. (2007). New applications of rapid prototyping and rapid manufacturing (RP/RM) technologies for space instrumentation. *Acta Astronautica, 61,* 352–359.
55. Romanides, J. S. (1965, April 8). All planets the same. *The Boston globe,* 18. Available online at https://blogs.ancientfaith.com/onbehalfofall/an-orthodox-perspective-on-alien-life, checked on January 3, 2017.
56. Rummel, J. D., & Billings, L. (2004). Issues in planetary protection: policy, protocol, and implementation. *Space Policy, 20,* 49–54.
57. Schwartz, J. S. J., & Milligan, T. (Eds.). (2016). *The ethics of space exploration.* Heidelberg, London and New York: Springer.
58. Schulte, W., Hofera, S., Hofmann, P., Thiele, H., von Heise-Rotenburg, R., Toporski, J., et al. (2007). Automated payload and instruments for astrobiology research developed and studied by German medium-sized space industry in cooperation with European academia. *Acta Astronautica, 60,* 966–973.
59. Sephton, M. A. (2014). Astrobiology can help space science, education, and the economy. *Space Policy, 30*(3), 146–148.
60. Siljeström, S., Lausmaa, J., Sjövall, P., Broman, C., Thiel, V., & Hode, T. (2009). Analysis of hopanes and steranes in single oil-bearing fluid inclusions using time-of-flight secondary ion mass spectrometry (ToF-SIMS). *Geobiology, 8*(1), 37–44.

61. Stephan, T. (2001). TOF-SIMS in cosmochemistry. *Planetary and Space Science, 49,* 859–906.
62. Sullivan, W. T., & Baross, J. A. (Eds.). (2007). *Planets and life: The emerging science of astrobiology.* Cambridge: Cambridge University Press.
63. Traphagan, J. (2015). *Extraterrestrial intelligence and human imagination: SETI at the intersection of science, religion, and culture.* Cham: Springer.
64. Traphagan, J. (2016). *Science, culture and the search for life on other worlds.* Cham: Springer.
65. UNOOSA, The official Report on the United Nations/Austria Symposium on Space Science and the United Nations. Source: www.unoosa.org/pdf/reports/ac105/AC105_1082E.pdf.
66. UNOOSA, The report on the United Nations/Austria/European Space Agency Symposium on Enhancing the Participation of Youth in Space Activities: Implementing the Recommendations of UNISPACE III. Source: www.unoosa.org/pdf/reports/ac105/AC105_774E.pdf (Section III E).
67. Vakoch, D. A., & Harrison, A. A. (Eds.). (2011). *Civilizations beyond earth: Extraterrestrial life and society.* New York: Berghahn Books.
68. Vakoch, D. A. (Ed.). (2013). *Astrobiology, history, and society: life beyond earth and the impact of discovery.* Berlin & Heidelberg: Springer.
69. Vakoch, D. A., & Dowd, M. F. (Eds.). (2015). *The drake equation: Estimating the prevalence of extraterrestrial life through the ages.* Cambridge: Cambridge University Press.
70. Waltemathe, M. (2017). A match made for heaven. Astrosociological and astrotheological aspects of spaceflight and religion. In *AIAA SPACE and Astronautics Forum and Exposition.* AIAA SPACE Forum, AIAA 2017-5157.
71. Weintraub, D. A. (2014). *Religions and extraterrestrial life. How will we deal with it.* Dordrecht: Springer Praxis Books.
72. Wilkinson, D. (2013). *Science, religion, and the search for extraterrestrial intelligence* (1st ed.). Oxford, U.K: Oxford University Press.
73. Willis, P. A., Greer, F., Fisher, A., Hodyss, R. P., Grunthaner, F., Jiao, H., et al. (2009). *Lab-on-a-chip instrument development for titan exploration.* Fall Meeting 2009: American Geophysical Union.
74. Workshop Report: Philosophical, Ethical and Theological Questions of Astrobiology, American Association for the Advancement of Science, Washington, D.C., 2007.

Printed in the United States
By Bookmasters